Randy Johnson

Manuals and Reports on Engineering Practice No. 73

QUALITY in the CONSTRUCTED PROJECT

A Guide for Owners, Designers and Constructors

Volume 1

Published by the
American Society of Civil Engineers
345 East 47th Street
New York, New York 10017-2398

First Edition (Revised version of the preliminary manual published in 1988)

ABSTRACT

This final version of the *Quality in the Constructed Project* (ASCE Manual No. 73) provides suggestions and recommendations to owners, design professionals, constructors and others on principles and procedures which have been effective in providing quality in the constructed project. It also provides guidance for establishing roles, responsibilities, relationships and limits of authority for project participants; and stresses the importance of concepts and practices that enhance the quality in the constructed project. Throughout the manual, various themes considered to be of particular importance are discussed. They include such concepts as: 1) the definition and assignment of responsibility; 2) the importance of teamwork; 3) the importance of concise contractual provisions; 4) the principles of good communication; 5) the owner's selection process for project team members; and 6) the procedures for design and construction.

Library of Congress Cataloging-in-Publication Data

American Society of Civil Engineers.
Quality in the constructed project: a guide for owners, designers, and constructors.
p. cm.—(ASCE manuals and reports of engineering practice; no. 73)
"Revised version of the preliminary manual published in 1988."
Includes index.
ISBN 0-87262-781-0
1. Civil engineering—Management—Handbooks, manuals, etc. 2. Building—Quality control—Handbooks, manauls, etc. 3. Civil engineering—Specifications—Handbooks, manuals, etc. I. Title. II. Series.
TA190.A54 1990
624'.068—dc20 90-14408
CIP

Library of Congress Catalog Card No: 90-14408
ISBN 0-87262-781-0
Manufactured in the United States of America.

MANUALS AND REPORTS ON ENGINEERING PRACTICE

(As developed by the ASCE Technical Procedures Committee, July 1930, and revised March 1935, February 1962, and April 1982.)

A manual or report in this series consists of an orderly presentation of facts on a particular subject, supplemented by an analysis of limitations and applications of these facts. It contains information useful to the average engineer in his everyday work, rather than the findings that may be useful only occasionally or rarely. It is not in any sense a "standard," however; nor is it so elementary or so conclusive as to provide a "rule of thumb" for nonengineers.

Furthermore, material in this series, in distinction from a paper (which expresses only one person's observations or opinions), is the work of a committtee or group selected to assemble and express information on a specific topic. As often as practicable the committee is under the general direction of one or more of the Technical Divisions and Councils, and the product evolved has been subjected to review by the Executive Committee of that Division or Council. As a step in the process of this review, proposed manuscripts are often brought before the members of the Technical Divisions and Councils for comment, which may serve as the basis for improvement. When published, each work shows the names of the committee by which it was compiled and indicates clearly the several processes through which it has passed in review, in order that its merit may be definitely understood.

In February 1962 (and revised in April 1982) the Board of Direction voted to establish:

> A series entitled "Manuals and Reports on Engineering Practice," to include the Manuals published and authorized to date, future Manuals of Professional Practice, and Reports on Engineering Practice. All such Manual or Report material of the Society would have been refereed in a manner approved by the Board Committee on Publications and would be bound, with applicable discussion, in books similar to past Manuals. Numbering would be consecutive and would be a continuation of present Manual numbers. In some cases or reports of joint committees, bypassing of Journal publication may be authorized.

AVAILABLE* MANUALS AND REPORTS OF ENGINEERING PRACTICE

Number	
10	Technical Procedure for City Surveys
13	Filtering Materials for Sewage Treatment Plants
14	Accommodation of Utility Plant Within the Rights-of-Way of Urban Streets and Highways
31	Design of Cylindrical Concrete Shell Roofs
33	Cost Control and Accounting for Civil Engineers
34	Definitions of Surveying and Associated Terms
35	A List of Translations of Foreign Literature on Hydraulics
36	Wastewater Treatment Plant Design
37	Design and Construction of Sanitary and Storm Sewers
40	Ground Water Management
41	Plastic Design in Steel—A Guide and Commentary
42	Design of Structures to Resist Nuclear Weapons Effects
44	Report on Highway and Bridge Surveys
45	Consulting Engineering—A Guide for the Engagement of Engineering Services
46	Report on Pipeline Location
47	Selected Abstracts on Structural Applications of Plastics
49	Urban Planning Guide
50	Report on Small Craft Harbors
51	Survey of Current Structural Research
52	Guide for the Design of Steel Transmission Towers
53	Criteria for Maintenance of Multilane Highways
54	Sedimentation Engineering
55	Guide to Employment Conditions for Civil Engineers
56	Subsurface Investigation for Design and Construction of Foundations of Buildings
57	Operation and Maintenance of Irrigation and Drainage Systems
58	Structural Analysis and Design of Nuclear Plant Facilities
59	Computer Pricing Practices
60	Gravity Sanitary Sewer Design and Construction
61	Introductory Manual on Computer Services
62	Existing Sewer Evaluation and Rehabilitation
63	Structural Plastics Design Manual
64	Manual on Engineering Surveying
65	Construction Cost Control
66	Structural Plastics Selection Manual
67	Wind Tunnel Model Studies of Buildings and Structures
68	Aeration—A Wastewater Treatment Process
69	Sulfide in Wastewater Collection and Treatment Systems
70	Evapotranspiration and Irrigation Water Requirements
71	Agricultural Salinity Assessment and Management
72	Design of Steel Transmission Structures
73	Quality in the Constructed Project—a Guide for Owners, Designers, and Constructors

*Numbers 1, 2, 3, 4, 5, 6, 7, 8, 9, 11, 12, 15, 16, 17, 18, 19, 20, 21, 22, 23, 24, 25, 26, 27, 28, 29, 30, 32, 38, 39, 43, and 48 are out of print.

Quality is never an accident. It is always the result of high intention, sincere effort, intelligent direction, and skillful execution. It represents the wise choice of many alternatives.

—*Anonymous*

The pursuit of quality is the cement holding the owner, design professional, and constructor together in a stable pattern where each supports the other in producing a successful project.

All project team members working cooperatively in an atmosphere of mutual trust and understanding:

- *Fulfill contract commitments competently and faithfully.*
- *Deal openly and honestly with all project participants.*
- *Protect the public health, welfare, and safety.*

Following these precepts the team members, acting with skill, integrity, and responsibility, reflect credit upon their individual professions and produce the desired quality in the constructed project.

PREFACE

This Guide has been written for all participants in a construction project, and describes a desirable *process* for project delivery from conception through design, construction, and operations start-up. It is a compendium of what the design and construction process should be to enhance quality. It contains descriptions of techniques, systems, methods, and procedures as contributed by numerous authors experienced in the process. It is not all-inclusive, and other options of equal or superior merit not mentioned in the Guide may exist or may be developed.

The Guide discusses numerous aspects of the process likely to be pertinent for major projects; for smaller projects, some of these aspects may need only limited attention or may not apply at all. Likewise, a description of multiple staff *functions* for a project is not to be taken as a need for multiple staff *positions*, because, for many smaller projects, the functions often can be accomplished by a single individual.

The Guide is intended to be educational in nature, with the belief that embracing the philosophies and processes it describes will contribute to the quality of a project. It is not, however, a complete codification of practice within the construction industry, nor does it represent a "baseline" or minimum standard for correct or appropriate project development. Rather, it is intended as an aspirational document. The authors and editors had the benefit of a variety of resources, including printed materials and the comments of several hundred reviewers. The attempt was made to select from these sources and present factors contributing to quality in design and construction, with the hope of stimulating readers to identify areas where the levels of their practice can be raised.

The Guide should be used with care, since there is no satisfactory substitute for the exercise of prudent judgment by the owner, designer, and constructor. Moreover, the specific contractual provisions involved in a project may vary the procedures suggested in this Guide and, in that case, the specific contractual provisions govern.

Finally, the Guide will be a living document, subject to ongoing review, coordinated by an oversight committee, and to revision at regularly scheduled intervals. You are invited to become part of this process by addressing your comments to the Manager, Professional Services, ASCE, 345 East 47th Street, New York, New York 10017.

ACKNOWLEDGMENTS

We are deeply indebted to the following individuals who have guided the development of this document over the past five years, and to the authors and numerous participants in the review process.

Steering Committee Members:
James W. Poirot, Chairman
Jerome S.B. Iffland
Harold L. Loyd
Robert L. Morris
John D. Stevenson

Implementation Committee Members:
Stephen C. Mitchell, Chairman
A.C. Burkhalter
Donald H. Kline
Robert A. Perreault, Jr.
Dean E. Stephan

ASCE Presidents:
Richard W. Karn (1984–85)
Robert D. Bay (1985–86)
Dan B. Barge, Jr. (1986–87)
Albert A. Grant (1987–88)
William J. Carroll (1988–89)
John A. Focht, Jr. (1989–90)
James E. Sawyer (1990–91)

ASCE Staff:
Edward O. Pfrang, Executive Director
R. Lawrence Whipple, Managing Director
Professional Affairs

Editors:
Leslie A. Clayton, Technical Editor
Harry Ferguson, Managing Editor (1986–88)
Grace M. Waldvogel, Managing Editor (1988–)

CONTENTS

Executive Summary xxi
Introduction xxi
Project Organization and Definition xxi
Selection of Design Professional and Negotiation of Agreement for Professional Services xxi
Alternative Investigations, Project Definition, and Design Criteria xxii
Preparation of Design, Plans, Specifications, and other Construction Contract Documents xxii
Preparing for Construction; Selecting Constructor; Construction Contract xxiii
Construction Activities; Construction Contract Submittals; Contract Administration xxiii
Conclusion xxiv

Chapter 1: Introduction and Definitions 1
1.1 Purpose 1
1.2 Principal Themes 1
1.3 Quality—Definition and Characteristics 1
1.4 Composition of Project Team 2
1.5 Establishing Requirements for Quality 2
1.6 Obligations of Team Members 3
Conclusion 3

Chapter 2: Owner's Role, Expectations, and Requirements 5
Introduction 5
2.1 Owner's Role and Responsibilities 5
2.2 Owner's Expectations and Requirements 5
2.3 Owner's General Requirements 5
2.4 Requirements of Private Owners 6
2.5 Requirements of Public Owners 6
2.6 Expectations vs. Requirements 6
2.7 Conflicting Expectations 6
2.8 Communication with Team Members 6
Conclusion 7

Chapter 3: Project Team 8
Introduction 8
3.1 Project Team Organization 8
3.2 Project Manager 8
3.3 Owner and Owner's Team 10
3.4 Design Professional and Design Team 11
3.5 Constructor and Construction Team 11
3.6 Alignment of Interest Among Team Members 12
Conclusion 13

Chapter 4: Coordination and Communication Process 14
Introduction 14
4.1 Importance of Coordination and Communication 14
4.2 Key Contacts of Team Members 14
4.3 Team Members' Roles in Coordination 14
4.4 Development of Coordination Process 15

Page

		Page
4.5	Communication Methods for Coordination	16
	4.5.1 Communication	16
	4.5.2 Forms of Communication	16
	4.5.3 Precautions	17
	4.5.4 Meetings	17
4.6	Critical Points in Project Communication	17
4.7	Timing	18
4.8	Resolving Conflicts and Disagreements	18
Conclusion		18
Chapter 5: Procedures for Selecting Design Professional		19
Introduction		19
5.1	Selecting Design Professional	19
5.2	Basis for Selection	19
5.3	Owner's Selection Committee	19
5.4	Statement of Qualifications	20
5.5	Selection Procedure	20
5.6	Advantages of Selection by Qualifications	21
5.7	Bidding	21
5.8	Other Selection Procedures	22
Conclusion		22
Chapter 6: Agreement for Professional Services		23
Introduction		23
6.1	Purpose of Agreement	23
6.2	Elements of Agreement	23
	6.2.1 Range of Services	23
	6.2.2 Scope of Services	24
	6.2.3 Instruments of Services	24
	6.2.4 Owner's Responsibilities	24
	6.2.5 Compensation for Services	24
	6.2.6 Other Contract Provisions	24
6.3	Standard-Form Agreements	25
	6.3.1 Professional Societies or Industry Associations	25
	6.3.2 Governmental Agencies	25
	6.3.3 Owners, Design Professionals, and Constructors	25
6.4	Short-Form Agreements	25
Conclusion		25
Chapter 7: Alternative Studies and Project Impacts		27
Introduction		27
7.1	Refining Project Requirements	27
7.2	Investigating Alternative Solutions	27
7.3	Determining Project Impacts	28
7.4	How Public Influences Alternative Studies	29
7.5	Selection of Preferred Alternatives	29
Conclusion		30
Chapter 8: Planning and Managing Design		31
Introduction		31
8.1	Organizing for Design	31
8.2	Design Team Leader	31
8.3	Initiating Design	32

Page

8.4 Project Design Guidelines 32
8.5 Coordination and Communication During Design 32
8.6 Monitoring and Controlling Design Costs and Schedules 32
8.7 Avoiding Threats to Quality 33
Conclusion 33

Chapter 9: Design Discipline Coordination 35
Introduction 35
9.1 Levels of Disciplines 35
9.2 Project Requirements for Each Discipline 35
9.3 Participation by Professional Discipline Leader During Design 39
9.4 Participation by Design Professionals During Construction 39
Conclusion 39

Chapter 10: Design Practices 41
Introduction 41
10.1 Office Operation 41
10.1.1 General Management of Design Office 41
10.1.2 Organization for Projects 41
10.1.3 Office Procedures 41
10.1.4 Filing and Storing Documents 41
10.1.5 Reference Library 42
10.1.6 Drafting Practices 42
10.2 Relationships with Owner and Constructor 42
10.3 Design Activities and Requirements 42
10.3.1 Design Considerations 43
10.3.2 Design Reviews 43
10.3.3 Construction Costs 43
10.3.4 Constructability Reviews 44
10.3.5 Peer Reviews 44
10.4 Design-Related Quality Control 44
10.5 Compliance with Codes and Standards 44
10.6 Approvals and Permits by Regulatory Agencies 45
10.7 Grant Procedures 45
10.8 Design Responsibility 45
Conclusion 46

Chapter 11: Pre-Contract Planning for Construction 47
Introduction 47
11.1 Owner's Capabilities 47
11.2 Resources for Quality Construction 47
11.2.1 Financial Resources 47
11.2.2 Materials for Construction 47
11.2.3 Manufacturing Capability of Suppliers 48
11.2.4 Human Resources 48
11.3 Regulatory Agency Requirements 48
11.4 Site Development 49
11.5 Review of Design-Construction Alternatives 49
11.6 Contractual Arrangements 49
Conclusion 50

Chapter 12: Construction Team 51
Introduction 51

Page

12.1 Contractual Arrangement 51
12.2 Field Organization for Construction 51
12.2.1 Owner's Resident Project Representative 51
12.2.2 Constructor's Construction Supervisor 52
12.2.3 Design Professional's Construction Support Services 53
12.2.4 Regulatory Agency Participation 53
12.2.5 Advisers for Construction 53
12.3 Assembling Construction Team 54
Conclusion 54

Chapter 13: Procedures for Selecting Constructor 55
Introduction 55
13.1 Selection Procedures and Qualifications 55
13.2 Constructor Qualification 56
13.3 Selection by Competitive Bidding 56
13.3.1 Role of Design Professional in Competitive Bidding 56
13.3.2 Competitive Bidding for Public Work 56
13.3.3 Bidding Procedures for Public Work 57
13.3.3.1 Prior to Bid Opening 57
13.3.3.2 Bid Opening 57
13.3.3.3 After Bid Opening 57
13.3.3.4 Contract Award 58
13.3.4 Competitive Bidding for Private Work 58
13.4 Competitive Selection Procedures for Negotiated Contracts 58
13.5 Noncompetitive Selection for Negotiated Contracts 59
Conclusion 59

Chapter 14: Construction Contract 60
Introduction 60
14.1 Construction Contract Documents 60
14.2 Content of Construction Contract 60
14.3 Form of Construction Contract 60
14.4 Standardization of Construction Contracts 61
14.5 International Construction Contracts 61
Conclusion 61

Chapter 15: Planning and Managing Construction Activities 62
Introduction 62
15.1 Organization for Project Construction 62
15.2 Preconstruction Meetings 63
15.3 Construction Activities 63
15.3.1 Construction Schedule 63
15.3.2 Estimates and Cost Control 64
15.3.3 Construction Facilities and Services 64
15.3.4 Material Management 64
15.3.5 Work Force Management 64
15.3.6 Safety and First Aid 64
15.3.7 Other Activities 64
15.4 Coordination and Communication 64
Conclusion 65

Chapter 16: Construction Contract Submittals 66
Introduction 66

Page

16.1 General 66
16.2 Planning Submittals Process 66
16.3 Scheduling, Preparing, and Processing of Submittals 67
16.4 Contract Documentation 67
16.5 Initial Technical Documentation 68
16.6 Shop Drawings for Structural Components 68
16.7 Shop Drawings for Manufactured Structural Components 68
16.8 Shop Drawings for Mechanical and Electrical Components 69
16.9 Shop Drawings for Equipment 69
16.10 Placing Drawings for Concrete Reinforcing Steel 69
16.11 Shop Drawings for Temporary Construction 70
16.12 Results of Independent Testing 70
Conclusion 70

Chapter 17: Contract Administration Procedures for Construction 71
Introduction 71
17.1 Quality Commitments 71
17.1.1 Materials 71
17.1.2 Workmanship 72
17.1.3 Statistical Quality Assurance and Quality Control 72
17.1.4 Requests for Substitution 72
17.2 Cost Estimates and Payments 73
17.3 Methods of Payments 73
17.3.1 Unit-Price 73
17.3.2 Lump-Sum 74
17.3.3 Cost-Plus 74
17.4 Retainage 74
17.5 Liquidated Damages 75
17.6 Bonus Clauses 75
17.7 Change Orders 75
17.8 Nonconstructor Invoices 75
17.9 Construction Progress 76
17.10 Progress Reports 76
17.11 Communication—Correspondence and Records 77
17.11.1 Written Communication 77
17.11.2 Records 77
17.12 Certificates of Completion 77
Conclusion 77

Chapter 18: Start-up, Operation, and Maintenance 79
Introduction 79
18.1 Planning for O&M Input and Training 79
18.2 Design Phase 79
18.3 Construction Phase 80
18.4 Start-up Phase 81
18.4.1 Planning the Start-up Program 81
18.4.2 Start-up Activities 82
18.5 Operating Phase 82
Conclusion 82

Chapter 19: Quality Assurance/Quality Control Considerations 84
Introduction 84
19.1 Getting Program Started—Owner 84

Page

19.2	Design Professional's Program		84
	19.2.1	Procedures	84
	19.2.2	Project Phasing	85
	19.2.3	Design Reviews or Audits	86
	19.2.4	Quality Control During Bidding or Negotiating Process	86
19.3	Constructor's Program		87
	19.3.1	General Elements	87
	19.3.2	Contractual Requirements	87
	19.3.3	Project-Specific Requirements	87
Conclusion			88

Chapter 20: Project Quality Through Use of Computers			89
Introduction			89
20.1	Evaluating Functions		89
	20.1.1	Project Computer Systems	90
	20.1.2	Design Quality	90
	20.1.3	Hardware and Software Requirements	90
	20.1.4	Judging Software Results	90
	20.1.5	Generic Data; Data Retention and Retrieval	90
20.2	Specific Computer Considerations for Design		91
	20.2.1	Project Programming	91
	20.2.2	Conceptual Design	91
	20.2.3	Preliminary Design	91
	20.2.4	Final Design	92
	20.2.5	Reconciliation of As-Designed and As-Constructed Data	92
20.3	Computers in Construction—Administrative Use		92
	20.3.1	Corporate Accounting	93
	20.3.2	Project Management (Cost Control, Scheduling, Material Control, Contracting) and Project Administration	93
	20.3.3	Special Applications	94
Conclusion			94

Chapter 21: Peer Review			95
Introduction			95
What Is a Peer Review?			95
What Is a Peer Reviewer?			95
Purpose of Peer Review			95
21.1	General Background		95
21.2	Benefits of Peer Review		97
21.3	Types of Peer Review		97
	21.3.1	Organizational Peer Review	97
	21.3.2	Project Peer Review	97
		21.3.2.1 Project-Management Peer Review	97
		21.3.2.2 Project-Performance Peer Review	98
21.4	Elements of Peer Review		98
	21.4.1	Request for Peer Review	98
	21.4.2	Scope of Peer Review	98
	21.4.3	Selection of Reviewers	99
	21.4.4	Review of Documents and On-site Interviews	99
	21.4.5	Reports	99
	21.4.6	Subsequent Actions	100
21.5	Responsibilities of Parties		100

21.6 Recognized Peer-Review Programs 100
Conclusion 101

Chapter 22: Risks, Liabilities, Conflicts 102
Introduction 102
22.1 Project Risks 102
22.2 Performance of Project Team Members 102
22.2.1 Qualifying Project 102
22.2.2 Qualifying Team Members 103
22.2.3 Contractual Arrangements 103
22.2.4 Performance Under Contract 103
22.3 Bonds, Warranties, and Insurance 104
22.3.1 Bonds 104
22.3.2 Warranties 104
22.3.3 Insurance 104
22.3.3.1 Insurance Needs of Owner 104
22.3.3.2 Insurance Needs of Design Professional 104
22.3.3.3 Insurance Needs of Constructor 105
22.4 Conflict Avoidance 105
22.5 Conflict Resolution 105
22.6 Litigation as Last Resort 106
Conclusion 106

Glossary 107
Acronyms 111
Appendices 113
1. Brief History of Development of Quality in the Constructed Project 113
2. Representative List of Standard-Form Agreements 117
3. Recommended Competitive Bidding Procedures for Construction Projects 123
Index 147

EXECUTIVE SUMMARY

INTRODUCTION

The purpose of this *Guide to Quality in the Constructed Project* is to provide suggestions and recommendations to owners, design professionals, constructors, and others on principles and procedures which have been effective in achieving quality in the constructed project; to provide guidance for establishing roles, responsibilities, relationships, and limits of authority for project participants; and to stress the importance of concepts and practices which may help achieve quality in the constructed project.

For the purposes of this Guide the project team members are defined as the owner, design professional, and constructor acting as independent participants. Traditionally, the design professional and the constructor are each bound to the owner by independent contract—the owner/design professional agreement for services, and the owner/constructor contract for the construction of project facilities. The design professional and the constructor have no contractual agreement, and all lines of their contractual responsibility run to and through the owner. Each team member's responsibilities are designated under the appropriate contract. Many other contractual arrangements for primary team members are possible. A one-contract arrangement has the owner contracting with a design-construct contractor, and a no-contract arrangement has owners accomplishing the design and construction of the project with their own employees. Regardless of the contractual arrangement used, the functions of owner, designer, and constructor are discrete and may be discussed as if independent parties were performing these functions.

Quality in the constructed project is defined as meeting the requirements of the owner, design professional, and constructor as specified by contract, while complying with laws, codes, standards, regulatory rules, and other matters of public policy. Thus, a fourth "partner," the regulatory agency, sits at the table with the three primary participants.

The owner, as originator of the project, is responsible for selecting the other team members, stating project requirements which meet the needs of the design professional and constructor as well as those of the owner, negotiating appropriate contracts, and for performing as team leader in a cooperative relationship with other participants in achieving quality in the constructed project.

PROJECT ORGANIZATION AND DEFINITION

The owner is responsible for: providing project financing; clearly stating project requirements; defining organization of the project team; setting up lines of coordination and communication; selection of other team members and contracting for professional design services and construction of project facilities; and for operating the project after construction is completed. In accomplishing these tasks the owner may find it necessary to seek advice on legal, insurance, financial, real estate, and other matters, in addition to assistance required in the planning, design, and construction of the project.

In stating project requirements, in structuring the project team, and in organizing communications, the owner may find it useful to employ the design professional at the inception of the project, in order to benefit from his or her expertise in all phases of the project, i.e., definition, planning, design, and construction. The owner may, with or without the advice of specialists, set the general criteria for engagement of the design professional. After the selection of the design professional, and during and after negotiation of the owner/design professional agreement for professional services, the project requirements can be rewritten from the general requirements initially stated by the owner to the more detailed project-specific requirements meeting the needs of the owner, design professional, constructor, and regulatory agencies for the particular project.

SELECTION OF DESIGN PROFESSIONAL AND NEGOTIATION OF AGREEMENT FOR PROFESSIONAL SERVICES

This Guide recommends selection of the design professional on the basis of project-specific qualifications following these steps:

- The owner structures a procedure for requesting and evaluating statements of qualifications from design professionals interested in providing services for the project.
- The owner, acting under a previously announced procedure, receives and evaluates qualifications of design professionals.
- The owner, having evaluated qualifications, solicits proposals from a short list of design professionals, selects the design professional, and negotiates the owner/design professional agreement for professional services.

The negotiation of this agreement defines the roles and responsibilities of each party, the requirements of the project, the scope of services required of the design professional, compensation for services, project budget and schedule, and other contractual matters. Agreements are negotiated by individuals who reach consensus on various elements of the relationship between the owner and design professional. Since the individuals negotiating this agreement may or may not be those who will actually engage in the activities to be performed under the agreement, it is necessary to express in writing the understandings between the parties. Clear communication is enhanced through the use of standard-form agreements which provide an initial framework. It is generally useful to have a legal review of contract terms and language.

ALTERNATIVE INVESTIGATIONS, PROJECT DEFINITION, AND DESIGN CRITERIA

After the agreement for professional services has been signed, the owner and design professional, working together under the terms of the agreement,

- refine and amplify previous written statements of project requirements;
- formulate and study alternative methods and arrangements for meeting project requirements;
- select the most favorable alternative;
- complete project conceptualization and planning;
- develop preliminary facility layouts and other project design criteria; and
- document this activity in the form of a written statement or report which can be used to guide the final design effort.

Under the terms of the owner/design professional agreement it is usual for the design professional to have the responsibility for making investigations, performing studies, accomplishing project planning, preparing reports, and other tasks under the broad direction of the owner. The owner has the responsibility to review and approve documentation presented by the design professional.

PREPARATION OF DESIGN, PLANS, SPECIFICATIONS, AND OTHER CONSTRUCTION CONTRACT DOCUMENTS

The design professional, acting under the terms of the owner/design professional agreement, is usually responsible for producing the completed design for the owner's approval. This effort is documented by plans and specifications and other construction contract documents used in project bidding and award for the owner/constructor contract. After these documents have been completed by the design professional they are presented to the owner and the owner's legal advisers for review and approval.

The design professional follows the preliminary design report approved by the owner in planning and executing the design effort, and is primarily responsible for design phase activities such as:

- planning and managing the design;
- coordination and communication during the design phase;
- monitoring and controlling design costs and schedule;
- providing professionally qualified staff;
- performing design-related quality control functions;
- designing in compliance with codes and standards, laws and regulations, and regulatory agency requirements; and
- arranging for appropriate design reviews, constructability reviews, operability and maintainability reviews, and peer reviews.

In addition to responsibilities under the owner/design professional agreement, the design professional also bears responsibility to protect the public health, safety, and welfare under state licensing laws, and to conform to the codes of ethics of the design profession.

PREPARING FOR CONSTRUCTION; SELECTING CONSTRUCTOR; CONSTRUCTION CONTRACT

In planning for project construction, the owner evaluates the financial, materials, and human resources available to the project. The results of these evaluations help to define the owner's contracting strategy and guide the owner in determining the field organization for construction of project facilities.

As the design phase is completed and project activities point toward the construction phase, the central activity becomes the selection of the constructor. The selection of a qualified constructor who performs at a competitive cost is an owner's decision that significantly influences quality in the constructed project. Procedures for selecting the constructor vary from highly structured public bidding procedures required for public agency contractors, to priced proposals, to selection of the constructor based on past associations with the owner.

The most important step in this process is the presentation and evaluation of constructor qualifications demonstrating the ability to perform under the conditions defined by the contract. Competition on the basis of qualification may result in a negotiated contract with the owner; it may result in a select list of constructors invited to bid on the project; or it may result in a prequalified list of constructors invited to bid on public projects.

The role of the design professional in the selection of the constructor lies in preparing, for the owner's approval, the bidding package, including contract documents which define the project and the procedures for submitting competitive bids or proposals. The design professional assists the owner in administering the bidding process, in evaluating bids or proposals received, and in awarding contracts.

The construction contract documents generally include the owner/constructor contract, general and supplementary conditions, plans and specifications, addenda issued before bid closing, constructor's bid, notice of award, performance and payment bonds, and contract change orders issued during project construction. These documents form the basis of understanding between the owner and constructor. The constructor is responsible for performing in accordance with the terms of the contract and for constructing the facility described in these documents.

CONSTRUCTION ACTIVITIES; CONSTRUCTION CONTRACT SUBMITTALS; CONTRACT ADMINISTRATION

In managing construction activities, the constructor is responsible for specific compliance with the owner/constructor contract; planning and enforcement of site safety programs; means, methods, and sequencing of construction; management and payment of subcontractors and suppliers; quality control related to construction activities; and meeting applicable codes, permit requirements, and other public agency regulations pertaining to his or her operations. In many of these activities the constructor looks to the owner and the owner's resident project representative (RPR) for review and approval of his or her activities under the terms of the contract. The design professional, although having no contractual relationship with the constructor, is actively involved in construction activities under the terms of the owner/design professional agreement. The owner usually contracts with the design professional to provide these technical services and may also authorize the design professional to perform certain administrative duties, including authorization to act as the owner's RPR for field construction activities.

Under the terms of the contract documents, the constructor is required to submit information for the review and acceptance or approval of the owner and, when so designated by the owner, the design professional. Submittals required by the owner/constructor contract may include contract compliance documentation; schedules and cash-flow estimates; structure of lump-sum bid items; shop drawings for structural components; equipment shop drawings; mechanical and electrical component shop drawings; performance data for equipment assemblies; drawings for temporary construction; results of independent testing, and other documentation. Preparation of information required is the responsibility of the constructor, assisted by suppliers, equipment manufacturers, and subcontractors, including detailers and fabricators. The owner, generally assisted by the design professional, is responsible for the review and acceptance or rejection of the constructor's submittals.

Under the overall provisions of the contract documents, the owner and constructor each assign responsibilities to assisting participants and agree on procedures and communications designed for smooth flow of steps in the submittal process. Attention to detail and prompt execution of procedural steps by each participant is essential to schedule maintenance and achieving quality in the constructed project.

The RPR is responsible for the administration of the owner's contracts involving activities during the construction phase of the project. In discharging this responsibility, the RPR places emphasis on: maintaining quality of materials and workmanship; considering and taking timely action on the constructor's requests for modification of construction contract terms; maintaining current estimates and making timely payments under the terms of contracts and agreements; monitoring construction progress; and building the project record, using all forms of written communication. In these activities the owner is assisted, under contractual terms, by the design professional and constructor.

CONCLUSION

Quality in the constructed project is achieved by the project team members and all other project participants working together to reach common goals. A cooperative atmosphere, rather than an adversarial climate, is likely to produce desired results. Contracts and agreements between the owner/design professional and the owner/constructor are structured to align mutual interests regarding money, time, decision making and performance.

The contractual arrangements and project organization stressed in this Guide present the owner, design professional, and constructor as independent team members, each with an area of prime responsibility. The owner is the key member of the team defining the roles of the project team members in the owner/ design professional agreement and the owner/constructor contract. In general, the owner retains responsibility for project definition and organization; for financial and site acquisition arrangements; for administration of contracts; and for the operation and maintenance of the completed project. The design professional has prime responsibility in the planning and design; in the preparation of the construction contract, to include plans and specifications and other documents; and in providing services for the owner during the project construction phase, including construction observation and technical review of submittals prepared by the constructor. The constructor has prime responsibility for construction of project facilities as specified by the contract documents; for job-site safety; and for protection of public health, safety, and the environment as impacted by construction activities.

Each project has its own unique set of circumstances requiring careful structuring of the contract terms which define the roles and responsibilities of each team member and the manner in which other team members assist in the effort to reach the common goal—quality in the constructed project. The responsibility matrix presented herewith is intended to serve as an illustration (not a standard) of assignment of roles and responsibilities which may be agreed to under the owner/design professional agreement and the owner/constructor contract.

Matrix of Suggested Responsibilities

TASK	O	DP	C	PRINCIPAL REFERENCE
Assemble owner's advisers	P			Ch. 2
Establish project requirements	P	A*	A*	Ch. 2
Arrange for project financing	P			Ch. 2
Define structure and organization of project team	P			Ch. 3
Plan coordination and communication process	P	A*	A*	Ch. 4
Establish policy and procedure for selection of design professional	P			Ch. 5
Select design professional	P			Ch. 5
Negotiate and sign owner/design professional agreement	P+	P+		Ch. 6
Perform alternative design studies	R	P	A*	Ch. 7
Evaluate project impacts	R	P		Ch. 7
Select preferred alternatives	P	A		Ch. 7
Complete project planning and set design criteria	R	P	A*	Ch. 8
Assemble and manage a qualified, multidiscipline design team	R	P		Ch. 8

(continued)

Matrix of Suggested Responsibilities (Continued)

TASK	O	DP	C	PRINCIPAL REFERENCE
Coordinate design disciplines and perform final design	R	P		Ch. 9
Coordinate planning activities with appropriate regulatory agencies	P	A		Ch. 10
Obtain necessary approvals and permits from regulatory agencies	P	A		Ch. 12
Set organization of field construction team	P	A	A*	Ch. 12
Establish policy and procedure for selection of constructor	P	A		Ch. 13
Select constructor by competitive bidding or other means	P	A	A*	Ch. 13
Prepare plans, specifications and other construction contract documents	R+	P+		Ch. 14
Sign construction contract	P+		P+	Ch. 14
Organize for construction	R	A	P	Ch. 15
Initiate job safety and first aid program	R		P	Ch. 15
Submit plans for temporary construction	R	R	P	Ch. 16
Present required contract submittals, including shop drawings		A	P	Ch. 16
Administrative review of contract submittals	P	A	A	Ch. 16
Technical review of contract submittals	R	P	A	Ch. 16
Construct project facilities as specified by contract documents	R	R	P	Ch. 17
Administer constructor, design professional, and other contracts	P	A		Ch. 17
Make appropriate progress payments	P			Ch. 17
Prepare and negotiate contract change orders	P+	A	P+	Ch. 17
Plan and staff for project start-up, operation, and maintenance	P	A	A	Ch. 18
Administer QA/QC programs for design activities	R	P		Ch. 19
Administer QA/QC programs for construction activities	R	R	P	Ch. 19
Consider, evaluate, and use computers as appropriate for all phases of project	P	P	P	Ch. 20
Specify use of peer review as necessary	P	A	A	Ch. 21
Perform competently and on schedule under contract terms	P	P	P	
Seek to avoid conflicts; or resolve them short of litigation	P+	P+	P+	Ch. 22

O:	Owner	P:	Primary Responsibility
DP:	Design Professional	A:	Advising or Assisting
C:	Constructor	R:	Reviewing

*If design professional or constructor are not yet under contract, these services are supplied by qualified advisers.

+Review by legal counsel is indicated.

Quality in the constructed project results when the state of mind of those involved in the project places quality foremost. It requires a major communication effort during the entire process to keep all parties informed of the vital elements of the work and the concerns of the owner, design professional, and constructor. It also requires mutual understanding of those concerns and a realization that few, if any, construction jobs are without problems. Finally, it requires determination of all parties to resolve these problems equitably as they occur.

CHAPTER 1

INTRODUCTION AND DEFINITIONS

This *Guide to Quality in the Constructed Project* had its beginnings with a series of meetings held in 1983–1985. These meetings, attended by leaders of the design and construction industry, were convened for the purpose of discussing efforts to achieve quality in the constructed project. It was finally decided that the American Society of Civil Engineers (ASCE) should develop and publish a guide to achieving quality in the constructed project. Appendix 1 gives additional information on the background and development of this document.

1.1 PURPOSE

The purpose of this Guide is to provide suggestions and recommendations to owners, design professionals, constructors, and others who work with them on principles and procedures which have been effective in achieving quality in the constructed project; to provide guidance for establishing roles, responsibilities, relationships, and limits of authority for project participants; and to stress the importance of concepts and practices that enhance quality in the constructed project.

This Guide is intended for the use of owners, design professionals, and constructors. Project users, operations and maintenance personnel, inspectors, subcontractors, and vendors will also find it helpful. It may be useful for government officials, university professors and students, judges, legislators, and others in explaining the design and construction process. The Guide is not in itself a technical document. Its language, style, and content are intended for non-industry readers as well as for trade professionals.

It is not the purpose of this Guide to set forth a minimum standard for correct or appropriate project development. Rather, it is intended to be an aspirational document, stimulating readers to identify areas where the levels of their practice can be raised.

It is not to be inferred that the procedures discussed herein will automatically result in quality in the project. Many other factors, some beyond the control of the project team, can affect the outcome. Depending on project characteristics, other procedures or modifications of procedures discussed herein may produce equally satisfactory results. The procedures outlined herein can and should be varied to meet unique project requirements while maintaining quality as an overriding goal.

1.2 PRINCIPAL THEMES

In the following chapters a number of themes considered to be of particular importance are discussed:

- Definition and assignment of responsibilities.
- Importance of teamwork.
- Understanding of requirements and expectations.
- Importance of contract provisions defining the exceptions and obligations of the project team members.
- Principles of good communication.
- Owner's selection processes for project team members.
- Need for adequate scope, time, and liability protection.
- Procedures for design and construction.
- Organizational, management, and administrative practices.
- Conflict avoidance; value of mediation.
- Benefits of peer review.
- Participation of the design professional during construction and start-up.
- Construction contract submittals, including shop drawings.
- Standard form of agreements and other documents.

1.3 QUALITY—DEFINITION AND CHARACTERISTICS

For the purposes of this Guide, quality is defined as meeting established requirements. Quality in the constructed project is achieved if the completed project conforms to the stated requirements of the principal participants (owner, design professional, constructor) while conforming to applicable codes, safety requirements, and regulations.

Quality can be characterized as:

- Meeting the requirements of the owner as to: function and appearance; completion on time and within budget; life cycle costs; operability and maintainability; environmental, health, safety, and people impacts; and other features.
- Meeting the requirements of the design professional as to: defined scope; adequate budgets; reasonable schedules; timely decisions by owner; interesting work for the staff; realistic risk sharing; reasonable profit; a satisfied client, and a finished project which results in positive recognition and recommendations for future work.
- Meeting the requirements of the constructor as to: a well-defined set of plans, specifications, and other contract documents; a reasonable schedule; timely decisions by the owner and design professional; fair treatment; realistic risk sharing; reasonable profit; a satisfied owner; and positive recognition and recommendations for future work.
- Meeting the requirements of regulatory agencies as to: public health and safety; environmental considerations; protection of public property, including utilities; and conformance with applicable laws, regulations, codes, standards, and policies.

The probability of achieving quality in the constructed project will be enhanced by complete and open communication among participants, selection of qualified personnel for all phases of the project, and rapid resolution of misunderstandings, conflicts, and disagreements.

It is important to understand that, under the definition stated, quality is not necessarily measured by appearance, durable materials, or other physical yardsticks. Instead, quality, as used in this Guide, is simply meeting the expressed requirements of the three principal participants, whatever they may be.

Thus, a temporary, sheet-metal enhoused, pump station with low capital cost, high operating cost, short expected life, and esthetic deficiencies, may well be a quality project if it meets the expectations and requirements of the three principal participants. Conversely, a Taj Mahal with all its beauty and durable materials may not qualify as a quality project if its construction results in cost or schedule overruns, litigation, environmental controversy, or negative impact on public health and safety.

The concept of quality construction through a team effort is also a key part of the quality definition. If the owner, design professional, and constructor are to be truly motivated to produce a quality constructed project, benefits must accrue to all three. This Guide stresses that the expectations and requirements of all three have the best chance of being met under the teamwork approach to quality.

1.4 COMPOSITION OF PROJECT TEAM

For the purposes of this Guide, the usual project team consists of the owner, design professional, and constructor. The responsibilities, roles, and limits of authority of these participants as discussed in this Guide are defined and controlled by the owner/design professional agreement, the owner/constructor contract, and other relevant contract documents. The structure of the project team and appropriate roles and responsibilities are discussed in Chapter 3.

The owner, as the originator and provider of funds for the project, is responsible for selecting the other team members and for leading and directing the project team. The owner selects qualified team members, using procedures discussed and recommended in Chapters 5 and 13, and directs their performance through the negotiation and administration of the agreements and contracts noted previously.

The term "team" is used intentionally, to imply a cooperative rather than adversarial relationship among the project participants. When members of the project team are competent and work together, chances for quality are greatly increased. The three team members, individually and collectively, both control and are the beneficiaries of quality in the constructed project.

1.5 ESTABLISHING REQUIREMENTS FOR QUALITY

Establishing the project requirements for quality begins at the project inception. A careful balance between the owner's requirements of project cost and schedule, desired operating

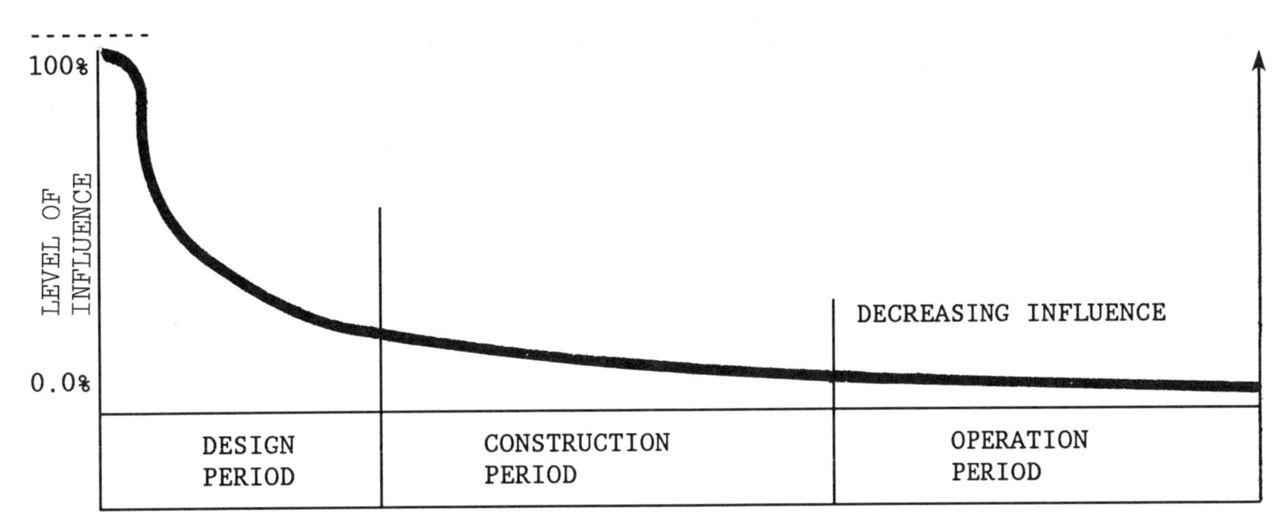

FIG. 1-1. Opportunity to Influence Project Characteristics

characteristics, materials of construction, etc., and the design professional's need for adequate time and budget to meet those requirements during the design process is essential. Owners balance their requirements against economic considerations and, in some cases, against chance of failure. The design professional is obligated to protect the public health and safety in the context of the final completed project. The constructor is responsible for the means, methods, techniques, sequences, and procedures of construction, as well as safety precautions and programs during the construction process.

Figure 1-1 illustrates how significantly the project characteristics and design are influenced early in the project. Decisions made in the initial stages of planning and design establish a program that is difficult and costly to change once construction begins. When the project requirements are clearly developed and communicated, competent design professionals retained, and adequate time and funding provided, optimum benefits of initial planning are realized.

1.6 OBLIGATIONS OF TEAM MEMBERS

All team members have an obligation to cooperate with one another to produce a quality constructed project meeting the conditions of applicable codes, standards, safety requirements, laws, and regulations. The design professional and constructor have additional obligations to accept only work that they are qualified to perform.

All three principal team members have the obligation to perform in an ethical manner and with integrity. A team effort will be difficult unless an atmosphere of mutual trust, respect, accommodation, and understanding prevails.

CONCLUSION

The purpose of this Guide is to present and discuss concepts and practices which may help achieve quality in the constructed project and to present suggestions and recommendations to owners, design professionals, and constructors which may help to achieve quality. This Guide is also intended for a broader audience, including other project participants and individuals seeking an understanding of the project design and construction process.

A listing of themes important to the project includes definition and assignment of responsibilities, principles of good communication, importance of teamwork, owner's selection processes for project team members, and other elements which help achieve quality in the constructed project.

Quality is defined as meeting the requirements of the owner, design professional, constructor, and, where appropriate, regulatory agencies. The requirements usually relate to project safety, costs, and schedules for all team members; to project functions, appearance, and operation for the owner; provision of a well-defined scope of services for the design professional; and a clear definition of responsibilities (contract documents) for the constructor. Other requirements include, for all team members, appropriate risk sharing, reasonable remuneration, timely decision making, good communications, and rapid resolution of misunderstandings, conflicts, and disagreements.

Quality in the constructed project is produced by the three team members—owner, design professional, and constructor—working amicably toward common goals.

CHAPTER 2

OWNER'S ROLE, EXPECTATIONS, AND REQUIREMENTS

INTRODUCTION

A successful project begins with the owner. The purpose of this chapter is to outline the broad responsibilities of the owner on most construction projects and to discuss two major factors that contribute to a successful quality project. The first is the development, by the owner, of complete and realistic expectations and requirements for the project. The second factor is thorough understanding by team members of the requirements, role, and responsibilities of the owner.

2.1 OWNER'S ROLE AND RESPONSIBILITIES

The owner is substantially responsible for the quality and success of any constructed project, by virtue of establishing the project requirements and communicating them to the other project team members. The owner is responsible for considering the relationships of cost and performance, appearance and function. The owner should understand the importance of such concepts and practices as life-cycle costing, peer review, alternative studies, value engineering, construction contract documents, and shop drawings. Owners who are not sufficiently knowledgeable in these areas should retain professionals with the appropriate knowledge and experience. The owner is well advised to refrain from being "penny-wise, pound-foolish" in such matters.

The owner is responsible for providing adequate financing for the entire project and making prompt payments to team members. The owner should be cooperative, encourage proper communications, and insist that all team members (owner's representatives, design professionals, and constructors) adhere to established project requirements.

These responsibilities apply to all types of construction projects and owners. However, depending on the type of project and the nature of the owner's organization, the owner may wish to delegate some responsibilities to other project team members. In other words, an owner's role can vary according to the specific owner, project, and contractual allocation of responsibilities and obligations. It is therefore essential that the owner and project team members thoroughly understand the owner's role and responsibilities before design and construction begin. For maximum clarity, it is recommended that each project team member's authority and responsibility be clearly assigned, agreed to, and expressed in writing.

In addition to the broad responsibilities stated in this section, the owner has other specific duties and responsibilities within the project team structure; they are addressed in the appropriate chapters. However, it must be emphasized that the contractual arrangements on any project are the primary source for defining these duties and responsibilities.

2.2 OWNER'S EXPECTATIONS AND REQUIREMENTS

As mentioned, owners' roles vary in accordance with the type of project, contractual arrangements, their organizational structures, and their capabilities. These variations influence the nature of their expectations and requirements for a specific project. A typical owner's expectations and requirements result from:

- The fundamental role of having or recognizing a need for a project.
- Past experience in executing projects.
- Observation or perception of what others have done to execute similar projects.
- Counsel and support services provided by advisers, including design professionals and other consultants.

2.3 OWNER'S GENERAL REQUIREMENTS

Most industry professionals would say that an owner's requirements are usually stated as: "I want a good job, on time, and within budget."

These generalized requirements usually can be achieved if: (1) The owner's definition of "good" is known, realistic, and completely understood; (2) the definition is communicated to and accepted by other team members; (3) the owner's time schedule and budget are realistic and accepted by the other team members; and (4) all project participants properly fulfill their roles.

Actual requirements for a particular project are more comprehensive. Requirements should be detailed and refer to specific aspects of a project, such as function, operation, schedule, technical matters, safety, quality, esthetics, fiscal, and administrative or management considerations. These requirements should be quantified to the extent feasible.

2.4 REQUIREMENTS OF PRIVATE OWNERS

The specific nature of requirements is influenced by the type of owner and the owner's role in a particular project. Private owners usually have fewer restraints and can therefore act more expeditiously than public owners. Private owners typically have stronger vested economic interests and are influenced by economic factors such as short- and long-term financing, size of the required investment, return on investment, profitability, and economic risk. Related aspects, such as demand, marketability, esthetics, and general fiscal performance, also influence the development of requirements by private owners.

2.5 REQUIREMENTS OF PUBLIC OWNERS

Public owners, such as local municipalities, state governments, school boards, utility districts, state or local development agencies, and the federal government usually have rigid processes and procedures under which they are expected to operate. Furthermore, if planning and funding a project consume an extended period of time, there may be changes in the funding agency's representatives, users, budgets, or programs, which may cause variations in roles and also adjustments in requirements. Public projects are often required to conform to established cost limitations, and their requirements tend to be performance and compliance oriented.

2.6 EXPECTATIONS VS. REQUIREMENTS

An owner's expectations exist even if they have not been expressed. A difference often exists between expectations and requirements. Satisfactory communication is necessary to understand the owner's expectations. Preliminary expectations should be thoroughly examined by the owner so that unrealistic ones can be detected and rejected. Expectations that are possible but may require additional effort, time, or cost can be evaluated. Once productive working relationships have been established, expectations are more likely to be revealed and fulfilled.

The design professional and the owner should develop a relationship that leads to identification of those expectations which are to be translated into written requirements. It is the stated, agreed-upon requirements by which quality in the constructed project is judged.

2.7 CONFLICTING EXPECTATIONS

Although the owner's expectations receive the most attention, all team members have expectations of some type. Conflicting expectations may exist, especially when numerous parties are involved. This situation demands sound communication and a system for stating and resolving differences among parties in a prompt and mutually satisfactory manner, so that quality in the project is not compromised or unreasonably disrupted.

2.8 COMMUNICATION WITH TEAM MEMBERS

Communication with the other project team members (the design professional and the constructor) regarding the owner's expectations and requirements for the project is most effective if they have assisted the owner in formulating them.

The continuing involvement of the design professional in all phases of the project can be achieved by early selection and utilization of his or her services first as an adviser in the conception and definition of the project, second as project designer, and third in representing and assisting the owner during the construction phase. In the case of the constructor, similar continuity of involvement is desirable. However, if competitive bidding is used, the option of continuing involvement with the constructor from concept to operation is not available to the owner. In this case the owner may elect to employ an adviser experienced in the construction phase of similar projects to provide the desired construction knowledge during the early project phases.

If the continuity of services described is not possible, a thorough understanding of the owner's requirements must be communicated to the design professional during the selection process and in the professional services contract, and to the constructor through the construction contract documents.

CONCLUSION

A successful project begins with the owner. Two major contributors to project quality are the owner's development and documentation of complete and realistic expectations and requirements for the project, and a thorough understanding among the other team members of the owner's role and responsibilities.

The key to fulfilling an owner's expectations and requirements is knowing and understanding what they are. Requirements are likely to be easily understood and quantifiable. On the other hand, expectations, though very important to the owner, are abstract and difficult to define and to understand.

A relationship where the design professional is working with the owner as an adviser during the project conception and definition phase is beneficial to both the designer and the owner, and is recommended. It allows the design professional to suggest various alternatives, estimate the order of magnitude costs, and identify trade-offs and other related aspects of the project. This function helps owners solidify their requirements.

An owner cannot expect poorly communicated requirements to be met. The owner and design professional should develop satisfactory communication and agree on how the requirements will be met and what expectations are reasonable. Thorough discussion of relevant facts, concerns, and necessities will enhance the probability that expectations will become requirements met.

The owner or his or her official representative should understand the importance of concepts and practices such as life-cycle costs, peer review, alternative studies, value engineering, construction contract documents, and shop drawings. The owner should be cooperative, plan for proper communications, require contract documents to be complete and accurate, and insist that construction adhere to project requirements.

CHAPTER 3
PROJECT TEAM

INTRODUCTION

Successful construction projects are conceived, planned, designed, and built by a project team consisting of an owner, design professional, and constructor. Quality is achieved when each team member's obligations are fulfilled competently and in a timely fashion, in cooperation with the other members. In most projects, each team member has functionally selected groups of individuals assisting in project activities. The organization and composition of these groups, as well as their attributes, activities, and responsibilities, are discussed in this chapter.

3.1 PROJECT TEAM ORGANIZATION

Figure 3-1 is a simplified organization chart illustrating the typical relationship between team members headed by the owner, design professional, and constructor. In this arrangement, each of the three principals is presumed to be an independent entity connected by contracts between the owner and design professional and between the owner and constructor. This is defined as the traditional project organization.

Many other organizational arrangements can and do exist with both public and private owners where divisions are not as clearly defined, and where two or more of the primary functions are performed by the same entity. Some arrangements are:

1. The owner employs his or her own design staff and contracts with the constructor only, or the owner may employ his or her own construction staff and contract with the design professional only.
2. The owner contracts with a design-construct firm.
3. The owner employs his or her own design staff and his or her own construction force.

Each of these arrangements can still be organized functionally along the lines shown in Figure 3-1.

3.2 PROJECT MANAGER

The term "project manager" as used here is a generic term for the individual who represents the owner and is responsible for the overall coordination and management of the project activities. The project manager may be a member of the owner's, design professional's, or constructor's staff, or may be an independent contractor employed by the owner, such as a professional construction manager.

For example, the project manager may:

- In the case of a public body, be the design professional, acting under the direction of the agency's board of directors.
- Be a member of the constructor's staff, when the owner has elected to contract with a design-construct firm to produce the project facilities.
- Be the owner him/herself, on smaller projects.

On some projects the owner may elect to contract with more than one design professional, to use multiple construction contracts, and to contract directly with suppliers. In this case a separate firm may be employed to perform the services of project manager. This service may be performed by a design professional, a constructor, a design-construct firm, or a professional construction management firm.

The project manager acts as the focal point of communications, and coordinates the project team's effort. Typical activities of the project manager include project initiation, project planning and scheduling, project start-up, administration of owner/design professional and owner/constructor contracts, communication and decision management, and project closeout.

Project planning and scheduling by the project manager involve developing a workable approach to the project and obtaining commitments from project team participants. During this activity, the project manager develops the owner's requirements into a formalized statement and expresses it to the other team members. The work of

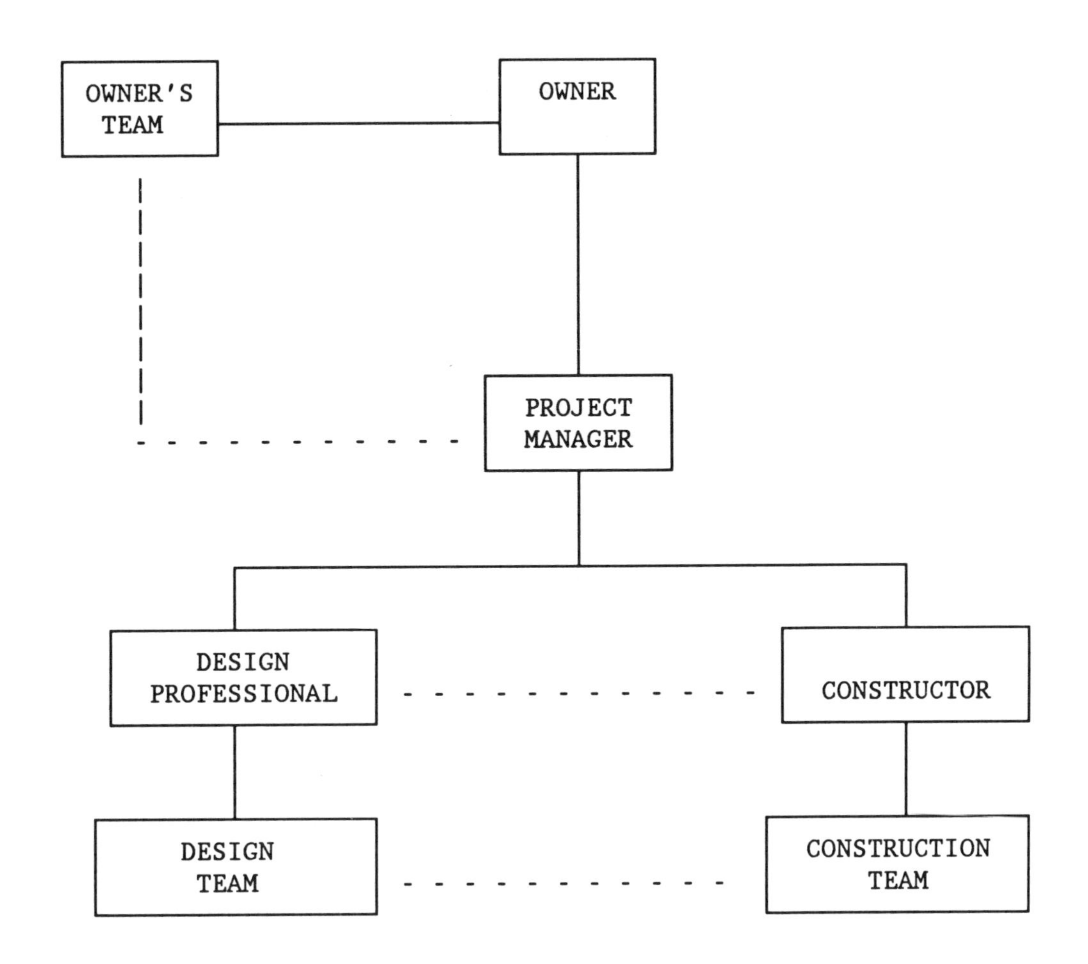

———— Lines of Authority (Defined by Contract)

- - - - - - - - Lines of Communication*

* Lines of communication between parties not bound to each other by contract are for the flow of routine information only and should be confirmed in writing.

FIG. 3-1. Project Organization Chart

each project team member is outlined. The project schedule is developed and confirmed, establishing milestones and deadlines. The project budget is reviewed and adjustments made. To summarize, the project manager sets the rules of the game.

With the planning complete, the design team is established and agreements are finalized. The design start-up usually begins with a pre-design meeting to establish policies and practices, requirements, expectations, schedule, budget and program, project data, quality control formats, and standards. Communication is critical at this stage, and every aspect of the project should be discussed and documented.

After design start-up, the project manager's role focuses primarily on coordination, communication, and administration of the overall effort. Progress meetings and design reviews are an effective way to communicate, check progress against schedule, evaluate design elements, monitor costs, coordinate activities, and enhance individual performance. The project manager is responsible for project leadership.

Qualities of a good project manager include an ability to:

- Formulate an effective management process and lead the project team to function under its precepts.
- Plan the project with respect to the level of performance required by the owner.
- Select an appropriate and qualified design professional to function on the project team.
- Facilitate and encourage good communications throughout the team for the duration of the project.
- Make and encourage sound and timely decisions and exercise strong leadership.

The project manager should serve the project from inception to completion. Continuity of the entire project team enhances effective communications, leadership, and greater responsiveness.

3.3 OWNER AND OWNER'S TEAM

As the initiator of the project, the owner assumes a primary role in establishing and directing the program to completion. The number and complexity of decisions to be made often make it necessary for the owner to assemble a group of advisers to function as part of the owner's team.

The function of the owner's team is to provide advice and assistance in areas such as financial and legal matters, user requirements, marketing studies and strategies, insurance, real estate, and other specialized fields. These advisers are generally bound to the owner by contract or as employees. They report to the owner but should be available for discussion with the project manager, design professional, and constructor during design and construction of the project.

As the initiator and primary moving force behind the project, the owner establishes the tone and level of performance required. The owner's responsibilities include articulating the particular need, funding, selecting the design professional and constructor, establishing project requirements, making decisions affecting project design and construction, and providing other essential information. Failure of the owner to fulfill these responsibilities can have serious consequences, regardless of the talent and abilities of the other team members. To a large degree, quality in design and construction depends on the performance of the owner.

The owner's responsibilities include—among others imposed by law, regulation, or the contract documents—the following:

- Fully disclose facts.
- Provide project funding.
- Enhance communications.
- Establish reasonable and attainable requirements.
- Delegate or assign decision-making appropriately and support that authority.
- Be realistic in assumption of risks and liabilities.
- Demand a quality project.
- Make timely decisions.
- Allow adequate time for good performance.
- Allow freedom for innovation.
- Exercise financial reponsibility and make timely payment.

Effective communication is a key element of a successful effort. Lack of communication about changes and delayed sharing of new information result in wasted time and money. Clear communication relating to relevant facts, requirements, expectations, scope of work, schedule, and funding is critical in any quality project.

Owners function most effectively within the project team when they have realistic expectations. Unattainable and/or unrealistic requirements discourage sound professional judgment and result in confusion within the team. Cost and time are two elements with which owners should exercise objectivity and realism in setting requirements.

Funding has a considerable impact on a project's quality, since the dollars support the process. Quality design takes time and costs money. Inadequate funding of the design effort cuts resources at a critical juncture of the project, where the resulting detriments are highly leveraged. Design costs are often on the order of one percent of the life-cycle cost of a project; yet the design is the single most important influence on project costs and quality.

The need for adequate time to perform necessary activities must not be overlooked by owners in favor of rushing a project from design to completion. A realistic and adequate schedule must be established to achieve the project's requirements.

When the owner delegates authority to a representative, this person is empowered to act on the owner's behalf, and the representative's decisions are supported by the owner.

3.4 DESIGN PROFESSIONAL AND DESIGN TEAM

The design professional's primary role is to conceive, plan, and provide quality design solutions in response to the owner's stated requirements. The design professional may be engaged by the owner to provide additional services through all phases of the project from initial investigations through design, construction, and start-up. If the design professional is not an employee of the owner, the owner/design professional relationship and responsibilities are governed by contract terms which delineate the responsibilities of each party.

Unless the design professional has a large, diversified and experienced staff, most projects are sufficiently complex to require that his or her staff be supplemented to provide additional qualified specialists. The design professional may add to the design team by subcontracting to geotechnical, electrical, mechanical or structural engineers, environmental planners, field surveyors, materials-testing laboratories, or others with specialized skills. Alternatively, some of these specialists may contract directly with the owner; but in any case the specialist functions as a member of the design team.

The following are typical responsibilities and functions of the design professional:

- Be qualified to provide services contractually undertaken and rendered.
- Apply appropriate skills to the design.
- Enhance communications.
- Achieve quality in design.
- Be responsive to established budget, schedule, and program.
- Make timely interpretations, evaluations, and decisions.
- Accept authority and responsibility.
- Cooperate and coordinate with efforts of others.
- Fully disclose related external interests.
- Avoid conflicts of interest.
- Comply with codes, regulations, and laws.
- Impartially interpret contract documents.
- Act with care and competence to represent the interests of the owner as required by the agreement.

However, whether some or any of these responsibilities/functions are to be within the responsibility of the design professional will necessarily depend upon the specific contractual allocation of services and responsibilities between the design professional and the owner.

The design team is formed with professionals whose experience and skills correspond to those needed for the project. The function of the design professional's team is to design for the owner a facility that meets project requirements and to provide plans and specifications and other contract documents from which the facility can be constructed.

Whereas the owner's role as the "initiator" is primary, the design professional's role is also important as the "implementor." The design professional's responsibilities and functions are central to achieving quality in construction.

The owner establishes the tone of the project and the level of performance desired, initiates and articulates the need, and contributes to the design. The design professional provides leadership, appropriate technical skills, and experience for the design effort. Both the owner and design professional contribute to achieving quality.

3.5 CONSTRUCTOR AND CONSTRUCTION TEAM

The constructor's role is to plan, manage, and accomplish the construction activities necessary to build the project in accordance with the plans and specifications and other contract documents prepared by the design professional. The contract for construction of the project is between the owner and constructor. The design professional is not a party to this contract, although the owner may elect to delegate certain functions to the design professional during the construction phase of the project.

The constructor's objective is to build a quality project safely and in compliance with the provisions of the owner/constructor contract. The constructor assembles a team of material and equipment suppliers, specialty subcontractors, material fabricators, and others to assist in the construction effort. These team members report directly to the constructor, who is responsible for them. The constructor's responsibilities include:

- Enhance communications.
- Build a quality project.
- Perform within the owner/constructor contract and approved change orders.
- Plan, implement, and assume responsibility for job safety.
- Make timely decisions.
- Be responsible for performance of subcontractors and suppliers.
- Provide skilled craftsmen.
- Coordinate and cooperate with other project team members.
- Comply with applicable codes, regulations, and laws.
- Act with care and competence.

The constructor is obligated to comply with the contract documents in the execution of work. In order to achieve quality in fulfilling that objective, the constructor maintains a program designed to provide quality workmanship and compliance with the contract documents (see Chapter 19).

3.6 ALIGNMENT OF INTEREST AMONG TEAM MEMBERS

The tone of this Guide implies harmonious cooperation among team members throughout the planning, design, and construction process. Differing interests among the team members may make this difficult to achieve.

The differences center on matters of:

Money: The owner seeks to contain costs, and the design professional and constructor want profit and prompt payment.

Time: The owner wants a compressed schedule, while the design professional and constructor need sufficient time to provide quality in design and construction.

Decision Making: The owner may delay decisions on design matters, thus interrupting efficient work flow for the design professional; the owner and design professional may delay decisions on construction matters to the disadvantage of the constructor.

Performance: The owner/design professional and the owner/constructor may fault each other's performance.

Uncontrolled self-interest and greed are likely to result in a project which exceeds the owner's reasonable budget and schedule, is not properly designed, is poorly constructed, and does not perform in accordance with anyone's expectations or requirements.

These problems need not occur if the project is approached reasonably by all team members with the common objective of building a quality project on time and within budget. This can be achieved if the following conditions prevail:

Money: The owner understands that "you get what you pay for," and the design professional and constructor achieve a profit commensurate with the risk and the required level of effort and skill.

Time: The owner understands the benefits of realistic scheduling to permit the design professional to investigate and study alternatives before arriving at design decisions, and to permit the construction to proceed efficiently following a logical sequential procedure.

Decision Making: All team members understand that the timely actions of each team member impact other team members and the project as a whole.

Performance: All team members understand that the performance of each member is most likely to contribute to a quality project if contracts are carefully thought out and the role and responsibility of each party clearly defined and executed.

If the interests of each team member are aligned by the items noted, it is possible to achieve harmony among team members, which can result in completion of a project with few surprises regarding schedule and budget, and the translation of competent design into quality construction.

CONCLUSION

Successful construction projects are conceived, planned, designed, and built by a project team consisting of an owner, design professional, and constructor. Within this context, quality is achieved when each team member's obligations are fulfilled competently and in a timely fashion, in cooperation with the other members. In most projects, each team member has functionally selected experts assisting in project activities.

The organizational arrangement integrating the roles of the owner, design professional, and constructor may be the traditional one where the owner contracts independently with the other two parties. In other arrangements, the owner may issue only one contract to a design-construct firm, or all of the functions may be performed within the owner's staff, with an outside design professional or constructor employed for unusual projects. Discussions in this Guide emphasize the traditional owner/design professional/constructor arrangement and outline the contractual requirements necessary to define this arrangement.

Owners are responsible to administer their contracts with the other project team members and to monitor and coordinate the activities of all parties involved in the planning, design, and construction of a project. The owner may discharge these responsibilities more effectively by delegating authority to a project manager.

In addition to the specific responsibilities listed, other responsibilities apply equally to all team members. These consist of accepting responsibility, striving for economy and efficiency, cooperating and coordinating with other team members, adhering to the established budget, schedule, and program, and insisting on quality.

In the structuring of contracts and the roles and responsibilities of individual team members, care must be taken to avoid adversarial relationships which interfere with the production of a quality project. With appropriate care and understanding, harmony among team members can result.

CHAPTER 4

COORDINATION AND COMMUNICATION PROCESS

INTRODUCTION

Recognizing that different types of constructed projects are properly organized, contractually undertaken, managed, and executed in different ways, this chapter provides general guidelines to establish an effective coordination and communication process essential to quality construction projects.

Coordination translates to effective implementation. For a project to be successful it should be viewed as a whole, and a plan developed and implemented that results in fitting the pieces together. Effective use of the skills and experience of all participants is necessary throughout the project and within the plan to achieve quality. Coordination requires effective and frequent communication among project team members.

Communication is the exchange of information. In any successful project it is essential that necessary information be communicated to the right individuals at the right time. This requires effective procedures, organization, and controls, and selection of team members committed to facilitating good communication.

4.1 IMPORTANCE OF COORDINATION AND COMMUNICATION

Effective coordination and communication tend to prevent problems which lead to the dissatisfaction of team members, even to project failures. Industrial owner studies of failures, near failures, and problems with newly constructed projects indicate that at least 25% of those failures resulted from poor communication or lack of coordination among the project team.

Insurance studies indicate that owners tend to resort to legal action not so much because of imperfections in a project, but rather because of unexpected events and surprises, mounting frustration over problems not addressed, absence of positive personal relationships or concern, or lack of information about problems. Frequency of lawsuits has been highest from clients with whom communications are difficult and those with limited construction experience.

In 1982, a subcommittee of the U.S. House of Representatives Committee on Science and Technology held hearings to examine the problem of structural failures in the United States and to identify factors that contribute most to such failures. Its report lists several factors significant in preventing structural failures, including:

- Good communications and organization in the construction industry.
- Timely dissemination of technical data.

4.2 KEY CONTACTS OF TEAM MEMBERS

For most projects, the focus of coordination and communication includes designated key contact persons for the owner, the design professional, and the constructor. Normally these key contacts include:

- The owner's project manager and/or resident project representative.
- The design professional's design team leader, and/or principal in charge.
- The constructor's supervisor and/or operations manager.

These key contacts have authority for coordination and communication in their areas of responsibility within the project organization.

Clear definition of project scope and of each participant's role in achieving it, together with proper controls to monitor project progress, cannot be overemphasized.

4.3 TEAM MEMBERS' ROLES IN COORDINATION

The roles of the project team members in the continuing coordination process change according to the particular phases of a project. Table 4-1 illustrates some of the primary roles of each team member during the project's beginning, design, construction, start-up, and completion.

The owner bears responsibility for selection of the project team members and for determining when each is to join the team. There is no single formula that is right for all situations, since many

TABLE 4-1. Roles of Team Members

Participant	Project Beginning	Design Phase	Construction Phase	Project Completion
Owner	Forming and informing the group. Leading in outlining project requirements for the design professional.	Contributing to decisions in support of design. Participating in design reviews. Communicating changes when necessary.	Providing for qualified inspection and testing as required by contract documents and regulatory agencies. Administering contracts.	Maintaining group coordination and getting the group's attention on follow-up or completion items.
Design Professional	Assisting with project objectives and program requirements. Leading the development of process for coordination among team members.	Leading the design effort. Involving the owner and others at appropriate times. Preparing necessary design plans and specifications.	Technical support for required interpretations, changes, shop drawing reviews, or field problems, in a timely way. Field observation.	Assisting with follow-up work, completing required manuals and documents, assisting with start-up.
Constructor	Being an early participant. Contributing to alternative studies and scheduling.*	Assisting in vendor selection and constructability reviews.*	Performing the construction effort. Involving others at appropriate times, such as shop drawings, inspections, tests, etc. Field observation.	Leading the follow-up. Guiding vendor and subcontractor follow-up work.

*Often the constructor is not yet selected for the project. If early selection is possible, the constructor can make these contributions.

complex factors contribute to making each project unique. Accordingly, the owner may be required to become involved in construction planning long before the constructor is selected, in order to properly coordinate the design and procurement phases of the project.

4.4 DEVELOPMENT OF COORDINATION PROCESS

Team members normally have expectations of each other throughout the life of a project. Expected characteristics include:

- Integrity, honesty, and trust.
- Open communication and dissemination of complete information.
- Competence in their respective roles.
- Commitment to the project's requirements.

The expectation of open communication and dissemination of complete information is difficult to fulfill, since it requires collective as well as individual actions. Project teams, like most other work groups, go through several stages as they evolve from "group" to "team," a process required for effective coordination. The stages are:

1. Developing an initial understanding at a personal level.
2. Developing a thorough understanding of the specific strengths and weaknesses of individual team members and their staffs.
3. Developing the team members' specific responsibilities for project requirements, work scope, procedures, schedules, and budgets, as well as the normal means of communication to be utilized.
4. Developing teamwork toward the project's requirements, with honest, clear, and timely communications, cooperative and coordinated actions, timely and consistent decisions,

prompt handling of problems, and keeping of commitments.

Establishing an atmosphere that encourages rather than inhibits open communication is essential to development of the coordination process. Problems that are not identified early and addressed appropriately will degenerate until it is no longer possible to correct, only to mitigate, the situation.

4.5 COMMUNICATION METHODS FOR COORDINATION

Communication among project team members is vital to achieving quality in construction. Particularly on large or complex projects, internal communications within individual team members' staffs are important. Whatever methods of communication are used to produce the required continuing coordination throughout the project, it is important that they be clear and frequent. Methods for coordination and communication during the project may include:

- Programming of the owner's requirements.
- Procedures established for the specific project.
- Use of schedules and updates on schedules.
- Use of budgets, cost studies, and alternative proposals.
- Meetings of the team, including subconsultants and subcontractors, when appropriate.
- Minutes or summaries of meetings, including identification of and responsibility for follow-up actions.
- Written contract clarifications.
- Discussions of team members with documented decisions or conclusions.
- Memos and letters with appropriate distribution.
- Reviews with transmittal letters.
- Progress reports, written or oral, or both.
- Joint reviews of documents, models, budgets, and schedules.
- Results and reports of field or lab tests.
- Joint visits to the site, vendors, fabrication shops, and test facilities.
- Formal reports of noncompliance/discrepancy/and other design or construction problems.
- Contract change orders.

Team leaders for the design professional and constructor can design feedback loops into the project's process to determine if needed communication is occurring.

4.5.1 Communication

Communication among team members is a highly complex process requiring effort and skill. When speaking, or "sending," it is important to distinguish clearly among: (1) Giving project-related information and objective data; (2) revealing concerns, opinions, feelings, or subjective data; and (3) initiating actions by way of requests, requirements, commitments, and changes.

When listening, or "receiving," it is important to pay particular attention to words and phrases, ideas, concepts and mental pictures, and feelings, whether spoken or implied.

Listeners should be certain that they understand that they are: (1) Receiving project-related information and objective data; (2) being advised of concerns, feelings, or subjective data; or (3) expected to initiate action as a result of the communication.

The most effective way to "receive" complete communication is to summarize understandings with the "sender" and correct any misperceptions immediately.

4.5.2 Forms of Communication

Effectiveness decreases from direct (face-to-face) communication to telephone, and again from telephone to written communication, depending on the level of detail or sophistication of what is to be communicated; but each form of communication has its place in continuing coordination. Generally, these guidelines apply:

- Direct communication during meetings or consultations is useful to define and address issues, problems, or complex matters, gather ideas interactively, and initiate important actions or decisions.
- Telephone conversations (or conferences) are useful to solicit information, provide sensitive information, or serve as urgent substitutes for direct communication.
- Written communications, such as memos, letters, or reports, are necessary to transmit factual information and to record decisions, agreements, and actions.

4.5.3 Precautions

It is important for all participants to recognize that the owner, design professional, and constructor have different backgrounds, qualifications, expertise, and expectations, as well as different definitions of a successful project. Other significant individual differences may also exist among team members and their staffs, including personal preferences for: Working alone or in groups; careful deliberation or prompt action; creativity or standard procedures; noticing detail or broad concepts; methods of handling disagreements; tactful or more direct approaches to dealing with people.

For coordination and communication to be effective, team leaders must understand and compensate for these individual differences. For example, a highly skilled design professional may be most effective working alone and may require reminders or encouragement to provide timely coordination with other design staff or with subconsultants.

As important as communication is, there is also a danger in overcommunication. If routine information is given wide distribution regardless of importance, the effect may be to camouflage issues which deserve more critical review. Instead of enhancing, this actually detracts from the project team's effectiveness because of the effort required of each recipient to evaluate the significance of the information received. A good balance is necessary to ensure adequate dissemination of needed information while avoiding overcommunication.

4.5.4 Meetings

Meetings are productive means of providing continuing coordination for projects. In order to be effective they should be brief and be conducted by a knowledgeable moderator who follows a prepared agenda. Project meetings are of two types:

- Regularly scheduled meetings (weekly, monthly), which generally follow a standard agenda and are used to track progress, identify problems and resolve routine conflicts.
- Special meetings which are called when needed to address particular situations, consider specific problems and develop unique solutions.

The following guidelines will help to make meetings more useful:

- Call meetings only to facilitate direct communication and problem solving.
- Explain the purpose of the meeting and clearly define each agenda item as informational, or needing discussion and/or action.
- Set time estimates or targets for each item on the agenda, as well as for the entire meeting. Keep presentations and discussions moving.
- Encourage everyone to contribute. However, stress that contributions should be relevant.
- Chart pads on easels are recommended for recording important points of discussion, summarizing decisions, and noting responsibilities, dates, etc.
- Seek to integrate ideas and settle conflict during discussions, to prevent the meeting from becoming disorganized and ineffective.
- Critique the meeting. Was it worth the time? What can be done to improve the format? What was helpful, what was not? What should be changed next time?
- Prepare accurate written summaries of actions, agreements, conclusions, and responsibilities resulting from the meeting, and communicate them to all attendees shortly after the meeting has taken place.

4.6 CRITICAL POINTS IN PROJECT COMMUNICATION

The following are examples of times or situations when coordination and communication among the project team members are especially critical:

- At definition of project scope, budget, and schedule.
- At definition of performance and quality criteria (refinement of scope).
- When concluding alternative or feasibility studies (affecting scope).
- When reviewing construction contract document language and requirements.
- When assessing economic or scheduling impacts of requested or necessary changes in scope.
- At substantial completion of design phase or phases.
- When evaluating constructor's or sup-

plier's suggestions for alternative methods, materials, or equipment.
- In unexpected situations that require changes in scheduled dates, scopes, procedures, costs, or materials.
- When dealing with significant problems of design or construction.
- At substantial completion of construction phase or phases.

When team members are not appropriately involved in the project or consulted about problems that directly concern them, they may develop negative, uncooperative attitudes which are not conducive to achieving quality in a project.

4.7 TIMING

Intervals between contacts among team members should be short, since people are more likely to complete tasks on time when contacts are frequent. This will prompt team members to handle the inevitable overloads or other problems more quickly. Frequent contacts also provide backup for other communication opportunities.

When team members receive key items late, such as a changed budget, schedule, or scope, they often react negatively, and schedules or quality of work may be jeopardized.

4.8 RESOLVING CONFLICTS AND DISAGREEMENTS

Regardless of team members' competence or the level of coordination, conflicts and disagreements may occur among them during the project. The following suggestions are offered for resolving disagreements among team members:

- Handle disagreements as soon as possible. Postponing may lead to hardening of opposing positions.
- Identify the project requirement that is the focus of the disagreement. This will help the team to avoid irrelevancies.
- Isolate the key issues in the disagreement. Starting with easier issues, address one issue at a time until each is discussed and resolved.
- Discuss all relevant facts and feelings on the issue before attempting to resolve it. Gather all the data. Attempting problem solving too quickly can lead to escalation and/or confusion.
- Develop several alternatives for resolution.
- If possible, come to team consensus on what is to be done. Forced solutions often lead to dissatisfaction. If a consensus cannot be reached, the owner, after consulting with other team members, should select the preferred alternative.

Indecision can be damaging, particularly if it causes disruption in the otherwise smooth continuity of a project. It is therefore important for each manager in the project organization to utilize his or her skills and authority to resolve conflicts prudently but quickly at the lowest possible level.

CONCLUSION

A well-coordinated effort among the various team members is required to achieve an integrated program which, in turn, is necessary to complete a quality project. Coordination requires effective and frequent communication among project team members and their staffs.

Insufficient coordination and communication have contributed heavily to project failures and problems and to the dissatisfaction of team members. Insurance studies indicate that owners tend to resort to legal action not so much because of imperfections in a project, but rather because of unexpected events and surprises, mounting frustration over problems not addressed, absence of positive personal relationships or concern, or lack of information about problems. The frequency of lawsuits has been highest from clients with whom communications are difficult and those with limited project experience.

Five essential elements comprise communication in any form:

1. What is communicated?
2. To whom is it addressed or intended?
3. What is intended to be understood by those who receive it?
4. What is actually understood by those who receive it?
5. What action is expected and why?

When the parties involved in the execution of a project establish a framework of dialogue that facilitates good communication, serious gaps are avoided between intended and actual understanding of project essentials. Attention to these essential elements assists in producing a project that meets the requirements of the owner, the design professional, and the constructor.

CHAPTER 5

PROCEDURES FOR SELECTING DESIGN PROFESSIONAL

INTRODUCTION

The selection and engagement of the design professional is one of the more important steps towards achieving a quality constructed project. It is therefore critical to the project that the owner carefully structure and administer a selection procedure that results in a proper fit between the experience, qualifications, and abilities of the design professional on the one hand, and the requirements of the project, on the other.

It is important that both the owner and the design professional begin their relationship on the project with an attitude focusing on quality in performance at reasonable cost, rather than placing an undue emphasis on lowest possible design costs. Disappointment may occur if design cost becomes the primary basis for selecting the design professional. Owners are not well served when they are parsimonious with resources spent on design services (approximately 5-10% of construction costs) which influence initial construction costs (100%) and lifetime operation and maintenance costs (200-500%).

5.1 SELECTING DESIGN PROFESSIONAL

No two professional design organizations represent the same training, experience, skills, capabilities, personnel, work loads, and particular attributes relating to design projects. Selecting the proper design professionals enhances the project team's ability to achieve quality in the constructed project. Sections 5.2 through 5.5 present recommended procedures for selection of the design professional on the basis of qualifications to meet project requirements.

5.2 BASIS FOR SELECTION

The owner establishes administrative policy and criteria for the selection of a qualified design professional. The owner's first step is to define the project requirements. In some cases this may be a general statement of the performance requirements of the project. At other times, the tasks to be performed may be individually identified and defined. The owner should also consider the need for professional design services during construction and start-up phases. By clearly defining through a request for proposals the services the design professional is to furnish, the owner can better evaluate which design professional is best qualified to provide the necessary services.

Factors considered in evaluating a design firm's qualifications include:

- The professional and ethical reputation of the design professional, as determined by inquiries with previous clients and other references.
- Professional registration of the principals and other responsible members of the design professional's organization in their state of residence, and registration or qualification to obtain registration in the state in which the project is to be located.
- The design professional's demonstrated qualifications and capability in performing the specific design services required for the project, including knowledge of codes or other governmental regulations.
- Evidence that the design professional has the necessary financial resources and business background to accept the assignment and provide full, continuous service.
- The design professional's ability to assign appropriately qualified staff to the project and ability to complete the required services.

5.3 OWNER'S SELECTION COMMITTEE

For larger projects, the owner usually designates a committee to select or recommend selection of a design professional. One procedure generally considered satisfactory makes use of a selection committee of three or more individuals, at least one of whom should be a professional engineer or architect. This committee is responsible for making recommendations after conducting appropriate investigations, interviews, and inquiries.

Others from the owner's organization who are often involved in the selection procedure are:

- The person who will be the owner's liaison with the design professional.
- The person who will be responsible for operating and maintaining the proposed project.
- The person who is allowed to make subjective judgments about the owner's preferences on esthetic elements, such as projects involving architecture or where image is involved.

The final choice of the design professional by the owner is based upon the committee's recommendations.

5.4 STATEMENT OF QUALIFICATIONS

In the selection procedure, the qualifications of the prospective design organization are the primary factor. Written qualifications submitted by design organizations are evaluated by the selection committee.

Many large industrial corporations, practically all branches of the federal government, and many state, county, and municipal agencies which engage design professionals maintain a file of "statements of qualifications" for firms engaged in various types of professional design services. If the owner does not maintain such a file, obtaining this information should be an early step in any selection procedure.

Such data are often presented on a form, in brochures, or by a combination of both. Federal agencies and many others use U.S. government forms SF254 and SF255 as a standard for data required in a statement of qualifications.

5.5 SELECTION PROCEDURE

The selection procedure is considerably enhanced when the owner is fully familiar with the purpose and nature of the proposed project, can describe the project and its requirements in detail, and can prepare a scope of services expected from the design professional. In some cases, the owner may not have a professional design staff available to define clearly the scope of work and describe the required services. In larger and more complex projects, an owner often retains an independent design professional to assist with these activities. The owner should still be familiar enough with the project requirements to communicate what is expected of the design professional. The selection procedure, however, can be modified to suit the circumstances.

The usual steps in the selection procedure are presented next. If the owner has had satisfactory experience with one or more design professionals in the past, it may not be necessary to follow all the steps outlined.

1. By invitation or by public notice, state the general nature of the proposed project, and request statements of qualifications and experience from design professionals or organizations who appear capable of meeting the project's requirements. Names and locations of consulting design firms may be obtained from the Professional Services Directory of *Civil Engineering* magazine, published by the American Society of Civil Engineers (ASCE), or from professional directories, such as those published by the American Institute of Architects (AIA), the American Consulting Engineers Council (ACEC), the National Society of Professional Engineers (NSPE), and other appropriate societies, or through the records of owners of similar projects.

2. Consider the statements of qualifications received. Select at least three design professionals that appear best qualified for the specific project. References from previous clients and performance of firms selected should be reviewed. It should be noted that often more than three organizations may appear to be qualified, in which case more firms may be considered. In fairness to those not finally selected, and due to the cost and time required to prepare and evaluate competent proposals, it is recommended that no more than five firms be considered.

3. In a letter of invitation to each design professional, the owner describes the proposed project in as much detail as possible, and includes a scope of work and outline of services required. The owner invites a proposal from each design professional, describing a plan for performing the work, the key personnel to be assigned, the schedule planned for completion and other appropriate information, such as location where the work will be performed, and financial capability. Each design professional is given an opportunity to visit the site, to review pertinent data previously compiled, and to obtain clarification of answers to any outstanding questions.

4. On receipt of proposals, the owner invites the design professionals to meet with the selection committee for separate interviews and discussions. During the interviews, the project's requirements and the professional services

proposed are discussed. During each interview, the selection committee reviews carefully each design professional's qualifications and experience record, capability to complete the work within the time allotted, key personnel to be assigned to the project, and the availability of qualified personnel to complete the project. The committee should be satisfied that each design professional understands the project's requirements and how to address them.

5. Check carefully with recent clients of each design professional under consideration and determine the quality of performance on other projects. This check should not necessarily be limited to references given by the design professional.

6. Prioritize the list of design professionals in the order of their ranking, taking into account criteria such as location, reputation, experience, financial capability, size, available personnel, references, work load, and any other factors peculiar to the project being considered.

7. Invite the design professional considered best qualified to appear for a second presentation to discuss the project further and to negotiate a fair compensation for the services to be provided.

8. The compensation requested by the design professional should be evaluated in light of the owner's previous experience and the range of charges reported by other users of similar services. Fair compensation is vital to the success of the project so that the full expertise of the design professional can be utilized.

9. If agreement is not reached with the first design professional selected, the negotiations are terminated by written notice, and similar negotiations then follow with the second design professional, and the third, if necessary, until agreement is reached. All such negotiations are strictly confidential, and the compensation discussed with one firm is not revealed to another firm.

10. When agreement has been reached on scope of services, level of effort, compensation, and schedule, the owner and the selected design professional should formalize their agreement in a written contract conforming to the guidelines presented in Chapter 6.

5.6 ADVANTAGES OF SELECTION BY QUALIFICATIONS

Selection of a design professional by evaluation of project-specific qualifications, with a later negotiation of contract terms, offers the these advantages:

- Professional judgment, an essential element in quality design, is duly considered and evaluated.
- Special studies and analyses unique to the project are considered and, if appropriate, included in the scope of the services.
- A scope of services responsive to the project requirements and to the owner's schedule and budget is developed jointly by the owner and design professional.
- The design professional's participation in the construction phase and other activities such as right-of-way acquisition, equipment procurement, start-up, and preparation of operation and maintenance manuals can be agreed upon and included in the contract.

Developing the scope of the design professional's services before project schedules and budgets are finalized offers the owner the advantage of utilizing the capabilities, experience, and judgment of the selected design professional before contract terms are finalized. Owners engage design professionals because they need a designer's professional experience and judgment. This expertise is needed throughout the project; but its greatest value to the owner is often during the planning and development of project activities, before a contract is signed.

5.7 BIDDING

Price-bidding for the procurement of professional design services is recognized as being counterproductive by professional engineering, construction, and architectural societies and by state and federal legislative bodies. In fact, the federal government and many states have adopted laws that require their respective governmental agencies to procure professional services by a competitive process based on qualifications similar to those just described.

The selection of a design professional on the basis of a low bid (price or hours of effort) for a predetermined scope of services and contract provisions prepared by the owner is not advised for these reasons:

- The low bidder may not be qualified to perform the services.
- Contracting on the basis of a fixed scope with a fixed price does not provide for flexibility, creativity, and ac-

commodation during the planning and design phases of the project.

- A meeting of minds between owner and design professional regarding project requirements may be more difficult to achieve.
- Predetermined scope of services may not be explicit or comprehensive enough to contain all design services required by the owner—resulting in contract change orders.
- The owner's insistence on low design costs generally leads to higher-than-necessary construction costs because analyses of alternatives and other creative approaches are not specified in the contract.

Qualifications-based selection is considered more advantageous to an owner, but it is not the intent of this guide to imply that competitive bidding based on price for architectural or engineering services by design professionals is unethical or unprofessional.

5.8 OTHER SELECTION PROCEDURES

ASCE Manual No. 45, *Consulting Engineering: A Guide for the Engagement of Engineering Services*, discusses two other selection procedures which are sometimes used.

1. Level of effort contracts are for providing supplementary assistance to an owner's staff. The client selects the design professional on the basis of qualifications and negotiated contract provisions, including fee.

2. The two-envelope system involves submission of a technical proposal in one envelope and a price (or hours of effort) proposal in a second envelope. The second envelope is to be opened only after the design professional has been selected on the basis of his or her technical proposal. The second envelope is then opened and used for negotiation of contract terms. If both envelopes are opened concurrently, or only the second envelope is opened, then price becomes a prominent factor in the selection, and the procedure has all the disadvantages of price bidding.

The two-envelope system is therefore not recommended. The extensive added cost to prepare a comprehensive scope and price without benefit of negotiation are considerable, especially for the firms not selected. This can discourage some quality firms from participating in the process, thereby limiting the opportunity to obtain quality proposals or bids. It can also lead to increased costs to the industry.

CONCLUSION

Owners are best served when their selection procedures follow those used by federal and many state agencies in their procurement practices—as outlined in the Brooks and mini-Brooks laws. Under these recommended procedures design professionals submit statements of interest and qualifications in response to an owner's invitation and statement of requirements for a specific project. The responses are evaluated by the owner according to previously announced selection criteria. Often, an owner conducts personal interviews with the three design professionals who appear to be most qualified for the assignment.

After the design professional is selected on the basis of qualifications to meet project requirements, contract negotiations between the owner and design professional are initiated. During these negotiations, scope of services, schedule, compensation, and other contractual matters are defined, agreed upon, and documented in a written contract. If the owner and design professional are unable to reach agreement, then the negotiations are terminated and the owner initiates negotiations with the next most qualified design professional.

This selection procedure allows an owner to select the design professional best suited to fulfill the specific project requirements, provides an opportunity for the parties to develop cooperatively a project-specific scope and schedule of services and time schedule, and provides a compensation program that is tailored to the scope of services and time schedule and that is fair and agreeable to both parties.

The best agreement results from establishing a fee after the full scope of services is understood by all parties. This may require extensive discussions utilizing the experience and knowledge of the owner, the design professional, and their advisers.

CHAPTER 6

AGREEMENT FOR PROFESSIONAL SERVICES

INTRODUCTION

The final steps in the owner's selection of the design professional as outlined in Chapter 5 are the initial steps in the structuring of the owner/design professional agreement for performance of professional services on the project. The agreement for professional services reduces to writing the decisions already made jointly by the owner and design professional relating to scope of services, schedule, fee, and owner/designer division of responsibility. Other features of the agreement will need consideration and definition.

Allowing sufficient time to negotiate and define the agreement for professional services benefits both the owner and design professional. While this statement may seem to be so obvious as to be universally followed, misunderstandings and poor quality result from the unwillingness of one or both parties to the agreement to discuss the project and refine the "meeting of minds." It has been wisely said, "A contract that has not been negotiated is a contract that neither party understands." Moreover, a clear, written expression of the duties and responsibilities of each party fosters mutual trust, prevents misunderstandings, and aids materially in achieving quality in the constructed project. This process normally should be undertaken with the advice and assistance of legal counsel.

6.1 PURPOSE OF AGREEMENT

Agreements are negotiated by individuals who reach consensus on various elements of the project which include the services to be provided, the time frame in which they will be performed, and the responsibilities of each party in accomplishing the work. Since the individuals negotiating the agreement may or may not be those who will actually engage in the activities to be performed under the agreement, it is necessary to express in writing the understandings between the parties.

The agreement communicates the intent of the negotiators to other individuals who may be involved at some later date, such as the owner's employees, the project manager, and the design professional's staff. If key members of the negotiating teams, including the project manager, are involved in performing under the agreement during the entire life of the project, the probability of a harmonious, mutually beneficial, and productive working arrangement is enhanced.

The purpose of the written agreement for design services between the owner and design professional is to delineate the responsibilities of each party and to define clearly the services to be performed. The goal of clear communication may be achieved through the use of standard-form agreements (discussed in sections 6.3 and 6.4) which provide the initial framework for an agreement. In all but the simplest contracts, legal review of contract terms and language is important.

6.2 ELEMENTS OF AGREEMENT

Although some terms of the owner/design professional agreement for professional services apply to virtually all projects and are contained in standard-form agreements, certain elements of the agreement must be tailored specifically for each project.

6.2.1 Range of Services

The range of services for construction projects which may be provided under the agreement are well defined in ASCE Manual No. 45 and are classified as follows: (1) Study and report phase; (2) preliminary design phase; (3) final design phase; (4) bidding or negotiating phase; (5) construction phase; and (6) operation phase. Detailed discussion of services to be performed under each of these phases is presented in Manual No. 45 and Engineers Joint Contract Documents Committee (EJCDC) documents.

The agreement specifies the range of services required from the design professional. The advantage of continuity of personnel from inception to completion of the project is a valid consideration in the owner's selection of the range of design services.

6.2.2 Scope of Services

The scope of services is a project-specific grouping of professional services by tasks to be accomplished as mutually accepted by the owner and design professional from the range of services which may be offered by the design professional. The tasking is generally divided into appropriate phases as outlined in 6.2.1. Tasks are defined and correlated with services required, level of effort by the design professional, project time, schedule of instruments of services, and compensation.

Understanding is enhanced when each task is defined and considered separately during the negotiations between the owner and the design professional. The division of the total services to be performed into discrete tasks, together with consideration of constraints of schedule and budget for each, form the basis for agreement of the owner and design professional. Quality is enhanced by specificity and, to the degree appropriate, by quantifying the effort.

6.2.3 Instruments of Services

Under the agreement, the design professional is expected to produce certain documents during the development of the project. Among these are preliminary reports on project feasibility, opinions of probable costs, alternative investigations and studies, impact reports, preliminary designs and outline specifications, final design, construction contract documents—including plans and specifications, reports on construction activities, record drawings, and other documentation. Each of these documents is a result of one or more tasks defined in the scope of work and is scheduled for delivery to the owner on a mutually satisfactory time schedule.

6.2.4 Owner's Responsibilities

The owner shares with the design professional the responsibility and the obligation for on-time performance of assigned tasks. This may include providing existing information on the project, arranging for additional specialized information necessary for design (e.g., field surveys, geotechnical investigations), coordinating activities with other project team members, arranging for permits and approvals from regulatory agencies, making prompt decisions, paying fees promptly, and other activities influencing the design professional's ability to perform under the terms of the agreement. It is important to quality that these responsibilities be discussed and written into the agreement.

6.2.5 Compensation for Services

The agreement for professional services provides for payment to the design professional for services rendered. ASCE Manual No. 45 contains descriptions and discussions of various methods of payment for professional services:

- Per diem.
- Retainer.
- Salary costs times multiplier plus direct nonsalary expense.
- Cost plus fixed fee.
- Lump sum.
- Percentage of construction cost.

In general, cost-related or effort-related fees are appropriate where the services to be performed have not been or cannot be completely defined. Lump-sum fee arrangements are appropriate when the scope of services has been fixed by mutual consent of the parties.

Provision is generally made for timely partial payment of fees as services progress. Time and amount of payment are determined by:

- Effort expended and costs incurred on a monthly or other appropriate time interval.
- Completion of contract phases or other project milestones.
- Estimated partial completion of lump-sum related services as claimed by the design professional and approved by the owner.
- Any other method agreed upon by the parties to the contract.

Since schedule and budget are areas of the project where misunderstandings frequently occur, these contract provisions should be carefully tailored to specific project requirements and completely understood. Standard-form agreements (section 6.3) contain appropriate language for various methods of payment.

6.2.6 Other Contract Provisions

The previous sections discuss elements of the professional services agreement which are project-specific. Other elements of the agreement common to all projects include:

- Duration of agreement.
- Termination of agreement.
- Ownership and reuse of documents.
- Insurance requirements.
- Limits of liability.
- Procedure to amend the agreement.
- Methods of dispute resolution.
- Legal jurisdiction controlling agreement.
- Official representatives of owner/design professional.

6.3 STANDARD-FORM AGREEMENTS

Framing a suitable agreement for professional services is a demanding and complex exercise but one which lends itself to a degree of standardization. In order to reduce the time spent in framing and reviewing necessary individual contracts and to give guidance and assistance, standard-form contracts have been developed.

6.3.1 Professional Societies or Industry Associations

A number of professional societies have cooperated with each other and with construction industry associations in producing standard-form agreements for professional services for use by owners and design professionals, as well as complementary, coordinated commentaries, construction contracts, and other documents. Primary contributors to this effort have been EJCDC, the American Institute of Architects (AIA), and the Associated General Contractors of America (AGC). Representative lists of documents produced by these organizations are given in Appendix 2.

6.3.2 Governmental Agencies

Most governmental agencies at the federal and state level, and some at local levels have their own service agreements which they prefer to use. These agreements incorporate legal constraints and contracting policy governing these public sector owners. Unless a public agency repeatedly contracts for professional services, the standard agreement may include provisions which are not appropriate for the purpose intended. Careful review of the terms of the contract by the design professional and the agency staff, with the unique requirements of the proposed project in mind, will benefit the project and the parties signing the agreement.

6.3.3 Owners, Design Professionals, and Constructors

Private organizations that frequently engage professional services may also develop standard-form agreements. In general, these standard agreements will have been developed unilaterally by the party tendering the agreement. Careful review of the form, contract language, and contract terms is necessary to adapt the agreement for design services to the unique requirements of the project and to provide equitable treatment of both contracting parties.

6.4 SHORT-FORM AGREEMENTS

In cases where the project is of a routine nature, relatively small and simple, a well-drafted short-form agreement for professional services may be appropriate. Short-form agreements can include:

- Letter of understanding outlining essential elements of the agreement, generally proposed by one party and countersigned by the other.
- Short-form contract proposed by either party with preprinted contract provisions.
- Industrial purchase orders with standard format and preprinted contract provisions.

The use of short-form agreements is appealing to busy people; but there is the danger of incomplete project definition and lack of mutual agreement among the parties. The use of industrial purchase orders is hazardous to both parties, since these documents have been developed to cover purchases of all kinds, primarily goods and sometimes construction, and may include language relating to guarantees and indemnifications not appropriate for a design services agreement. These clauses may render the agreement uninsurable under the design professional's professional liability insurance while omitting worthwhile terms and conditions.

CONCLUSION

Agreements for professional services are tailored individually for each project, and document the agreements between owner and design professional which began during the final stages of the selection process and were fully developed with the negotiation of appropriate contract language to cover all phases of project development, design, and construction.

Elements of the agreement which demand detailed consideration and definition include the range of services to be provided, scope of services, instruments of services, compensation for services, and responsibilities of both the owner and design professional. Negotiating and framing the agreement for services on a project requires time and effort by the contracting parties and should be undertaken with the advice and assistance of legal counsel.

In order to expedite individual agreements and to provide a guide for structure and language, standard-form agreements are available from professional societies, industry associations, owners (public and private sectors), design professionals, and constructors. The standard-form agreements advanced by the professional societies (EJCDC and AIA) after consultation with industry associations (AGC and others) have the advantage of input from a number of sources and tend to be more objective and impartial than agreement forms developed unilaterally by one of the contracting parties.

The writing of a clear and complete document to communicate the agreement of the contracting parties is in the interests of both the owner and design professional and to the benefit of the project.

CHAPTER 7

ALTERNATIVE STUDIES AND PROJECT IMPACTS

INTRODUCTION

Conceptualizing and planning for construction projects requires development and study of various alternatives. These activities are a joint effort of the owner, design professional, and constructor (if available). The resources spent in formulating, investigating, and studying alternative approaches to decisions will vary depending on the size and complexity of the project.

Even the smallest projects are subject to alternative decisions related to site selection, schedule, materials and equipment, and many other elements. On small projects, alternatives may be studied and evaluated by the owner or the design professional in an unstructured way during the planning and design phases. On the larger, more complex projects a structured program of alternative study is generally required.

Project impact (physical, economic, social) analysis and reporting may be required by federal, state, or local laws. Such mandated analysis and reporting is a separate and distinct activity from alternative formulation and study for project enhancement, but these activities influence each other and should be conducted concurrently.

The study of alternatives and project impacts is in a sense an open ended activity generally conducted by the design professional under the terms of the owner/design professional agreement. The choice of study topics, extent of investigations, level of effort, reporting, and decision making should be carefully considered and defined in this contract. Flexibility to vary the level of effort assigned to foreseen alternatives and to consider new alternatives exposed during the progress of the design phase is desirable.

7.1 REFINING PROJECT REQUIREMENTS

During the early project phases which include conceptualization and preliminary planning, the stated project requirements are examined and revised to provide guidelines for design and construction. This examination considers answers to the following typical questions:

- What do the owner, design professional, constructor, and regulatory agency want and expect from the project? What is needed?
- Is there understanding and agreement on project requirements?
- What limitations are placed on this project?
- What aspirations or restrictions will guide the design and construction?
- Are project requirements achievable and realistic in view of constraints on budget and schedule, environmental compatibility, etc.?

Working together, the owner, design professional, and constructor (if part of the team at this stage) determine the answers to these and other equally pressing questions and prioritize project activities leading to design and construction.

At the end of this process the owner, with the assistance of the design professional, will have made most of the decisions pertaining to the general scope of the project, its expected community and political acceptance, its environmental impact, the geographic location, the functions of the facility, requirements and priorities during construction—including scheduling and installing equipment requiring long lead times, and the approximate cost.

7.2 INVESTIGATING ALTERNATIVE SOLUTIONS

The study scope and level of effort to achieve quality in design may vary considerably, depending on the nature and needs of the project. Various alternatives studied will affect project performance and appearance, life-cycle cost, cost-benefit ratio, schedule of completion, and socioeconomic and environmental impacts.

The number of alternatives chosen for examination, the extent to which each is subjected to detailed planning evaluation, and whether more than one "preferred" alternative is selected for final design study are key decisions best made early in the project planning and scoping process. These issues are usually resolved as a preliminary to the agreement between the owner and design professional, but changes during the study are not uncommon, and flexibility in the scope and terms of the agreement is recommended.

A broad view of project possibilities, based on owner, designer, and constructor experience, knowledge, innovation, and teamwork is essential to selecting the preferred alternatives for further study.

As a minimum, proposed alternatives should:

- Be responsive to the project requirements.
- Recognize legal requirements.
- Address any preestablished criteria agreed upon by the affected parties.
- Comply with land use and zoning regulations.
- Be functionally efficient, i.e., technically correct, cost-effective, and constructable; safe; and environmentally and esthetically acceptable.

Beyond these basic requirements, the alternatives can vary greatly in type and complexity.

Typical examples of alternative studies include:

- Scheduling alternatives, such as: a schedule selected by the constructor, owner, or both; rapid (compressed) schedule at extra cost but desirable because of short construction season or compensating advantage such as more rapid return on investment; delayed versus immediate construction start; seasonal construction; schedule imposed by a third party, such as a regulatory agency.
- Functional alternatives, such as: materials handling methods; traffic flow arrangements (patterns in air, water, land, people, or products); types of travel modes (vehicle type, size, style); methods to provide fish passage at barriers in waterways; space allocations, clear-span requirements in buildings; and public, private, or joint-use options for a facility.
- Conceptual planning or layout alternatives, such as: alternative route studies; site locations; drainage methods; structural systems and materials; and construction methods affecting design.
- Cost alternatives, including: design cost; capital cost of construction; operation and maintenance costs; various life-expectancy or design-life periods; return on investment; cost comparison of building the total facility now versus building some portions later; value of extra cost for esthetics; and cost/benefit ratios.

Alternative analysis cost studies during the early stages of project development are generally based on relative costs for segments or portions of the project, and not on total or absolute project costs. The purpose of such cost studies is to help make decisions on various project features. Early-stage estimates are classified as "order of magnitude" estimates and are made without specific, detailed engineering data. Actual costs of alternatives under consideration may vary appreciably from the costs used in the early alternative comparisons, but the relative costs are valid for selection purposes. For the owner's budgeting purposes, a more refined estimate is made after the best alternatives are selected and developed. A host of variables will later determine actual costs, i.e., current labor and material costs, market conditions (competition), site conditions, final scope of the project, time schedule, etc.

The number of early alternatives studied depends on the owner's objectives, the skill and innovative capacity of the design professional, the cost and complexity of the project, and the technological nature of projects designed to meet similar needs.

7.3 DETERMINING PROJECT IMPACTS

Each construction project has an impact on its environment in some way, depending on its size, function, physical characteristics, and the surrounding area. When developing alternatives, design professionals and owners consider how the proposed alternatives affect such features as: wetlands; aquatic and wildlife resources; farmland resources; scenic vistas; navigable waterways; natural stream or water-body quality; natural vegetation, including forests; cultural (historical and paleontological) resources; topographic features; or air quality. In addition to these environmental and cultural aspects, the proposed alternative's impacts on the social fabric of the community are considered (potential displacements, possible effects on nearby residential units, libraries, hospitals, schools, transportation lines, etc.). Deciding whether or how a project will impact a surrounding area is seldom solely the responsibility of the owner. Quasi-governmental bodies, such as planning and zoning boards, community boards, and various commissions often have considerable influence on determining a project's impact and status. Public opinion also plays a role in shaping a proposed alternative. This is discussed in section 7.4.

Determining a proposed project's impact on physical and socioeconomic environments is often a vital step in developing a quality project. Furthermore, considerable time and money can be saved when any necessary modifications are made as early in the design process as possible. The owner should retain competent and experienced professionals and allow enough time at the inception of a proposed project to study its potential impacts and develop reasonable alternatives.

A project's impact on its environment is often considered when permits of some sort are required before construction can begin. Besides routine building and occupancy permits, some projects may require special federal, state, or local approval or permits which are usually related to the proposed project's impact on the surrounding environment. Before granting a permit, an agency usually requires the owner to submit various types of documentation demonstrating that the proposed project's effect on the environment has been evaluated and that proper measures to control adverse effects will be exercised when the project is built and operated. Many states have specific environmental regulations that specify the type of documentation necessary to evaluate a proposed project. Most state regulations on these matters have been promulgated since passage of the National Environmental Policy Act (NEPA) in 1969, and state procedures usually adopt a NEPA-type approach to environmental evaluation. It is important that owners and planners understand that some projects or alternatives may be subject to an extensive approval process by federal, state, and local agencies. That process influences the development and selection of alternatives and the overall project.

Before a project is submitted for review or approval, the owner, with the assistance of the design professional and other advisers, determines what federal, state, or local laws and regulations apply to the proposed project and identifies the agencies responsible for administering them. This determination assists the owner in developing a strategy for early coordination with appropriate agencies and helps consolidate information these groups need to review the project.

7.4 HOW PUBLIC INFLUENCES ALTERNATIVE STUDIES

On occasion, a particular alternative will be subject to review and comment by members of the affected communities. The appropriate reviewing agency usually informs the public through a newspaper notice of a meeting during which the public can review the application. Public interest reviews have often influenced project designs and, in some cases, have contributed to delay and cancellation of projects if the public perceives adverse environmental results to be unacceptable. Since public reaction can influence the design of certain projects, owners should create a coordinated strategy to facilitate public awareness of key aspects of the proposed alternative. Communicating with the public and addressing its concerns in the planning process can forestall public demand to address new issues at later stages, such as during final design or construction.

7.5 SELECTION OF PREFERRED ALTERNATIVES

Alternatives for elements of the constructed project are studied and compared one to another by evaluating salient features such as utility, costs, esthetics, environmental compatibility, technical feasibility, constructability, or risk. A numerical rating system may be devised to give each feature a weighting in respect to other features. The sum of these numerical ratings is used to derive the preferred alternative.

The design professional generally performs the study of alternatives and presents an analysis of the relative merits of salient features to the owner. The owner, assisted by advisers, then selects the preferred alternative. Selected alternatives for the project as a whole or for elements of the project are structured to:

- Meet project requirements.
- Meet federal, state, and local laws and regulations.
- Provide mitigation measures for adverse impacts.

Consideration of project impacts and impact reporting, especially where public hearings and public information programs are mandated, may dictate the formulation and selection of alternatives not anticipated when project requirements were initially defined.

CONCLUSION

The study of alternative project elements by the design professional and the selection of the preferred project element by the owner, with the assistance of advisers, is an activity common to construction projects of all sizes. The definition and study of alternatives goes hand in hand with the refinement of statements of project requirements during the conceptualization and planning in the early stages of project development.

The choice of alternatives to be evaluated is influenced by the nature of project impacts on the surrounding environment. If a public information program is mandated by federal, state, or local regulations, alternatives are carefully chosen, studied, and selected with a view to presenting the facts of the study and alternative selection procedure to the general public or to the directors and/or staffs of public agencies. Particular emphasis is given to alternatives which provide mitigation measures for adverse impacts.

Planning, designing, and constructing a quality project usually involves the consideration of a reasonable number of viable alternatives. The formulation, study, and selection of various alternative project elements requires the participation of all members of the project team as well as participation of the appropriate regulatory agencies. The study and analysis of alternatives is generally delegated to the design professional with assistance, where appropriate and available, from the constructor. The results of the alternative studies are presented to the owner who, with the assistance of advisers, selects the preferred alternatives. The alternatives are structured and selected to meet overall project requirements, to conform to federal, state, and local regulations, and to recognize and mitigate adverse environmental impacts.

CHAPTER 8

PLANNING AND MANAGING DESIGN

INTRODUCTION

In the design phase of a project the functional requirements stated during the conceptual phase are given form, and documents are prepared to define the project for construction. Planning and managing the design effort involve elements of organization, staff selection, direction, control, and coordination essential to achieving quality in the project.

The complexity of the design effort varies with the individual project. For example, a design activity plan may be as simple as an outline of tasks to be performed (for a small, straightforward project) or may be a complex loaded flow chart that includes time and hour of effort requirements, all as appropriate for the project at hand. The following discussion describes steps in a process that can be condensed for small projects or expanded for large projects.

8.1 ORGANIZING FOR DESIGN

The design professional or his or her designated design team leader is responsible for executing the design phase from initial development to completion of plans, specifications, and other construction contract documents.

With a clear understanding of the project's requirements a plan can be prepared. This project design plan often takes the form of an arrow diagram or flow chart identifying the multiple activities required to deliver the project to the owner. The plan identifies the relationships among the various activities required to complete the project, noting the duration of each activity and perhaps the hours of effort needed to perform it. With a detailed project plan, milestones can be set and related to design costs. This information should reflect the scope of services as outlined in the contract, the contract documents agreed upon, and the schedule for professional services with the corresponding design budget. A detailed list of activities assists in identifying experienced team members and areas of expertise, as well as providing direction in establishing project files.

The design plan is the road map for successful completion. Deviations from the plan are recognized and corrected to keep the design effort on schedule and within the design budget. Corrective action may involve modification of the original schedule and budget.

8.2 DESIGN TEAM LEADER

The design team leader is the design professional or a designated member of the design professional's staff and is the key interface between the owner's project requirements and the development of design. The design team leader has the duties and responsibilities of:

- Developing the project's design budget, reflecting the resources and organization necessary to perform the work.
- Developing the specific design schedule within the overall time available for the entire project.
- Developing a project-staffing plan including the required expertise, and assembling the design team.
- Developing assignments for the design team.
- Developing checklists for contract documents and deadlines for completion of phases.
- Coordinating development of the project procedures.
- Monitoring and managing the design team's performance; making prompt decisions.
- Updating design agreements when required by scope changes, schedule delays, or other events.
- Scheduling in-house and owner reviews.

These responsibilities, coupled with the master project schedule, enable the team leader to monitor overall progress on the design effort and identify potential problems.

8.3 INITIATING DESIGN

All but very small project design efforts are initiated by a design team meeting during which the design team leader reviews with other design team members the project's scope and requirements. Task budgets are reviewed, and the relationship among various tasks is discussed. The need to meet all schedules while remaining within budget is emphasized. Commitment of the design staff to provide their services on schedule and within budget is verified.

Having the full commitment of the design team, the design team leader meets with the owner or project manager to review the delivery schedule for professional services. If the owner or project manager sees the need for a revision in the scheduled design milestones, these changes are reviewed with the design team members for possible conflict, and action is taken prior to initiation of design.

The design team's ability to meet schedule commitments and remain within budget are directly related to the owner's performance in providing on schedule those items specified in the owner/design professional agreement and providing timely review and comment on submittals made by the design team.

8.4 PROJECT DESIGN GUIDELINES

Each project has several sets of requirements that are communicated to or developed by the project design team. Sometimes the requirements are stated in numerical terms of performance. For example, a bridge should be designed to carry a specific loading, or a wastewater treatment plant should provide a certain level of effluent quality at a certain flow rate. At other times, an owner may only have a general idea of what the constructed facility should look like or how it should perform. In these cases, the design team leader, working with the team, first develops a set of project design guidelines to refine project requirements. This may involve preparing alternative studies (discussed in Chapter 7) to evaluate proposed concepts. The alternatives are reviewed with the owner to obtain agreement on the design approach that meets performance requirements within the project budget. Based on this agreement, the design team develops specific project design guidelines. It is extremely important to the achievement of quality that the owner and design professional reach early agreement on the expected results of the design and document these agreements in writing.

The proposed design solutions are structured to comply with federal, state, and local codes and regulations. In some cases, regulatory agencies will influence the investigation of alternatives and the design approach. Here it is important that discussions which include the owner, design professional, and regulatory agencies be initiated early in the design effort and continue through final design to avoid unnecessary delays or surprises in the agency-approval process.

8.5 COORDINATION AND COMMUNICATION DURING DESIGN

The design team leader keeps the owner and design team members informed on the design's status, normally submitting monthly (or more frequent, if necessary) progress reports to the owner. These reports contain information on meetings held and work accomplished during the reported month, and work to be accomplished in the following period. Design problems should be recognized as early in the process as possible; those that may require a change in scope, budget, or schedule should be identified and the difficulty resolved. If a different perspective on the project evolves, a revised design may be necessary.

Since most projects require more than one design discipline, design team meetings are held regularly. Such meetings are a means to familiarize other design team members with various aspects of the design process. Items discussed include schedules, budgets, and design coordination. Even though each member adheres to the project-design guidelines, conflicts are possible. These are discussed and resolved. If the solution conflicts with the owner's stated requirements, a meeting is arranged with the owner to explain the difference and to determine appropriate solutions.

8.6 MONITORING AND CONTROLLING DESIGN COSTS AND SCHEDULES

The design professional regularly monitors design-cost reports, which reflect budgeted versus actual expenditures. The expenditures should be consistent with the progress to date. If more time is being spent in developing a certain portion of a project than was budgeted, an explanation is found and compensating action taken. If additional services not specified in the contract are necessary to meet the overall project requirements, this information is communicated to the owner immediately, and amendment for additional services is negotiated.

A project schedule has a series of milestones and dates for submittals of reports identified in the contract. Adherence to these key dates by both parties is necessary to keep the project on schedule. Interim submittals keep the owner informed and able to offer comments early, when budgets and schedules can best accommodate change.

8.7 AVOIDING THREATS TO QUALITY

After a design professional has been selected or an in-house design team has been structured, threats to quality of the project can be avoided by:

- Developing the scope of services to meet project requirements.
- Developing the work plan for the design phase of the project.
- Estimating accurately the hours of effort required and cost involved to achieve a quality design.
- Recognizing that most programs are incomplete, that changes are inevitable, and that budgets and schedules usually need to be revised accordingly.
- Developing a realistic schedule.

It is important that associate consultants performing services under an agreement with the design professional contribute to the discussion of the scope of design services. If the consultant does not have an opportunity to assist in developing the scope of services, or is not consulted in a timely manner when problems develop which impact on his or her technical discipline, the quality of multidiscipline projects may suffer.

The design professional in private practice is confronted with owners who insist on contracting for professional services by means of competitive bidding to a fixed scope of services. When such services are procured on this basis, the scope of services is fixed prior to bids and usually without contribution from the design professional. When tasks are not discussed with and agreed to by the owner, the designer must reduce services to no more than what he or she unilaterally believes is absolutely necessary within budget limitations. Quality is severely threatened if selection is made on price without a thorough dialogue on qualifications and scope. Such a practice should be avoided.

Fast-track construction, where design and construction activities on the project run concurrently, may be perceived as best meeting overall project objectives, but often presents a threat to quality of design. Construction of portions of the project before final design is completed tends to freeze elements of design, thus precluding investigations of some alternatives. Changes in design criteria are difficult to accommodate. The added pressure and time constraints of having construction in progress further complicate review and checking procedures, inviting errors and making project communications more difficult.

Avoiding threats to quality includes assigning personnel to the design team who are experienced and knowledgeable in assigned tasks and who serve on a continuing basis until their functions are completed. Staff overloads and extreme overtime schedules should be avoided.

Avoiding threats to quality also includes assigning experienced design professionals to information-gathering and field survey teams. Personnel experienced in design considerations will investigate buried facilities on the project site, capacities of existing utility lines, or drainage facilities that must serve the project, as well as other site conditions that have an impact on the design and construction.

CONCLUSION

Under the owner/design professional agreement, the owner is responsible for furnishing agreed-upon information and other assistance on schedule. The owner monitors design activities, reviews and approves contract documents and other submittals, and provides prompt decisions on problems requiring owner input.

The design professional, through his or her design team leader, organizes the design effort, provides experienced and knowledgeable staff, develops design tasks, and monitors the performance of the design team. The design professional manages the design effort and communicates appropriately with the owner on matters affecting design progress, schedule, and budget.

After investigation and study of design alternatives, the design professional develops project design guidelines for the owner's review and approval. Design guidelines are also discussed with regulatory agencies' staffs and modified to meet agencies' requirements when appropriate.

Quality in the design of the project can be achieved only if threats to quality are avoided.

Positive actions to be taken by design professionals, their design team leaders, members of their staffs, and consultants, include realistic development of scopes, work plans, task assignments, hours of effort estimates, schedules, and budgets. Positive action to be taken by the owner includes the selection of a design professional on the basis of qualifications, providing a complete and realistic set of requirements, keeping informed on the project, showing willingness to make changes as new information or ideas become available, establishing a single point of contact with authority to act for the owner, and providing adequate and timely funding.

CHAPTER 9

DESIGN DISCIPLINE COORDINATION

INTRODUCTION

Most projects for design professionals involve more than one professional discipline. Three multidisciplinary types of projects are addressed in this chapter.

1. The engineering-design project, such as an industrial plant or a transit system.
2. The architectural-design project, such as an office building, commercial complex, or monumental structure.
3. The design-construct project which integrates both design and construction under a single responsibility.

Listing three types of projects by no means limits the number of project classifications or combinations. Other types of delivery systems exist, the form depending upon the design objective to be achieved.

The goal of each multidiscipline design team is to provide a facility that meets the project's requirements. Team members from each design discipline integrate their technical knowledge with that of the practitioners from other disciplines to satisfy the overall design objectives. With many projects some compromises in efficiency and cost may be necessary to meet these integrated objectives. Each discipline's effort considers that the safety of the user, the site workers, the public, and the environment are of primary importance.

The organizational examples presented as follows represent one concept of multidiscipline interrelationships for a project. Each project has a design professional responsible to the owner for the design. The interrelationship of the disciplines will vary with the contractual responsibilities that have been established for the project.

9.1 LEVELS OF DISCIPLINES

For multidiscipline projects, three organizational levels generally apply:

1. The design professional's design team leader has prime responsibility to the owner for meeting the project's design requirements and for staffing the design team with individuals or subconsultants who are experienced in the principal and support disciplines.

2. Principal discipline practitioners supply the technical expertise needed to plan and design an integral part of a project. They have the responsibility to design those portions of the facility covered by their specialty in coordination with other portions of the project. Examples are structural, mechanical, and electrical engineering specialties.

3. Support discipline practitioners supply the technical expertise needed to measure and test the physical characteristics of the project site and construction material. Examples are geotechnical investigations, materials testing, topographic surveys, and hydrologic analysis.

The design team consists of members from disciplines needed to complete the project. Each team member understands the project requirements, and applies his or her specialty to achieve a portion of the completed design under the direction and guidance of the design team leader.

The responsibility of each design discipline practitioner extends beyond the completion of the construction contract documents to the interpretation of design in the construction phase of the project, consistent with the contract for professional services.

9.2 PROJECT REQUIREMENTS FOR EACH DISCIPLINE

The requirements and responsibilities of the team members from each discipline involved in a project depend on the type of project proposed, the project's requirements, and the contractual relationships among the various parties. Three examples of projects that illustrate organizational concepts and general contractual relationships are: (1) The engineering project; (2) the architectural project; and (3) the design-construct project.

The objectives of an engineering project are usually controlled by the utilitarian or functional requirements of the constructed facility. Exam-

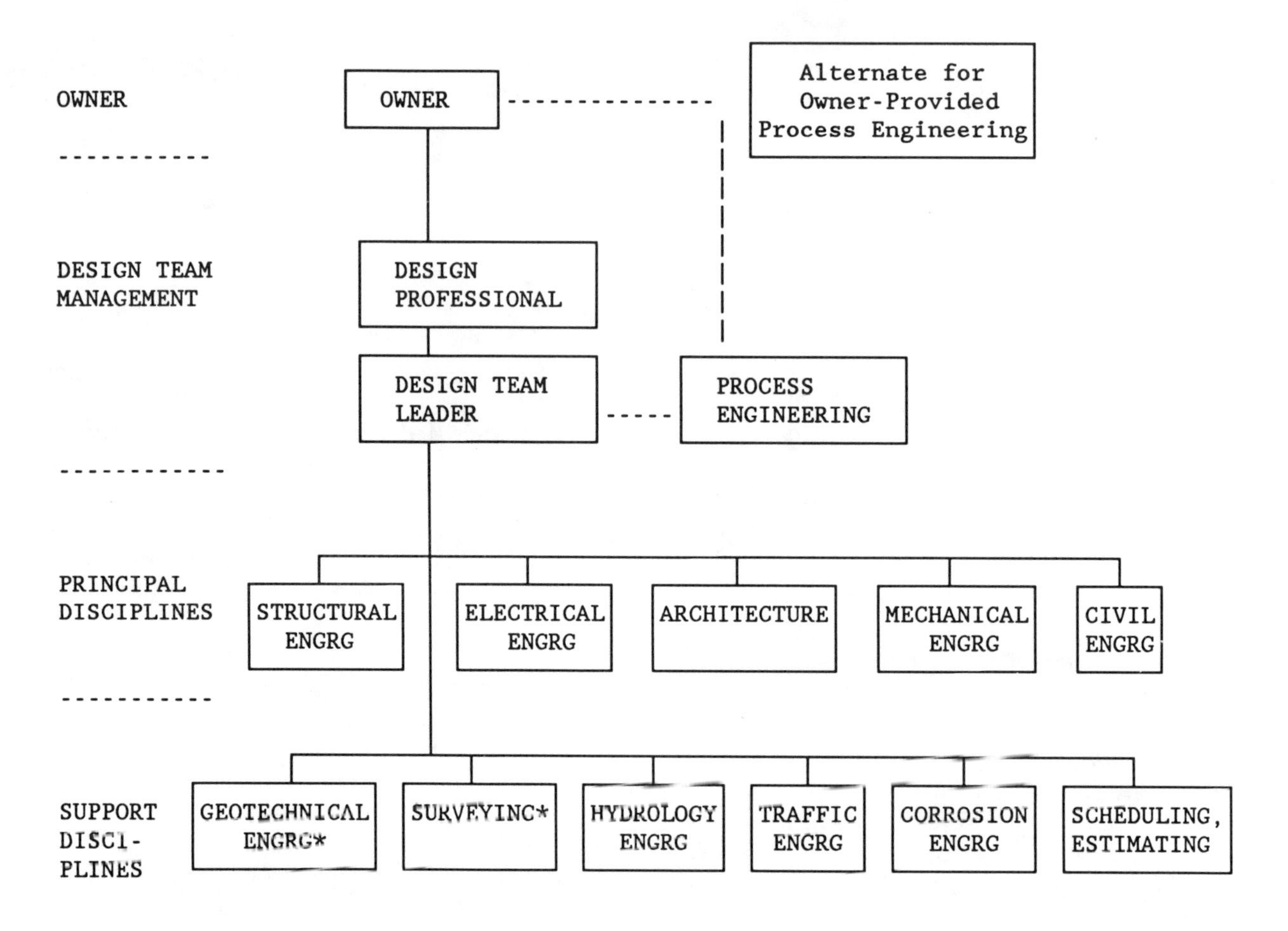

FIG. 9-1. Example of Multidiscipline Project Organization: Engineering Project

ples of engineering projects include industrial plants, transit systems, and wastewater treatment facilities. The design team leader for the engineering project usually has expertise directly applicable to the project. For example, mechanical engineers may lead a cement plant project, where the process flow determines the plan arrangement; structural engineers may lead a transit system project, where bridges or underground structures comprise the primary project components. Figure 9-1 illustrates these types of design organizations.

The requirements of an architectural project are primarily controlled by architectural rather than engineering considerations. Examples of architectural projects include office buildings, retail-commercial complexes, educational facilities, and monumental structures. Like the engineering project, the architectural project requires the contributions of many different design disciplines. The only difference in the design organization is the discipline that is designated as the lead or prime. For an architectural project, the prime discipline is architecture, and principal and support disciplines form the second and third design tiers, respectively. Figure 9-2 illustrates the architectural project design organization.

The design-construct project incorporates both design and construction under a single responsibility. The design-construct manager usually has two subsidiary managers—the design team leader and the construction team leader—who report to the design-construct manager. The design team organization below the design manager is the same for either an engineering or architectural project. Figure 9-3 illustrates a typical design-construct project organization.

Design team objectives are determined by

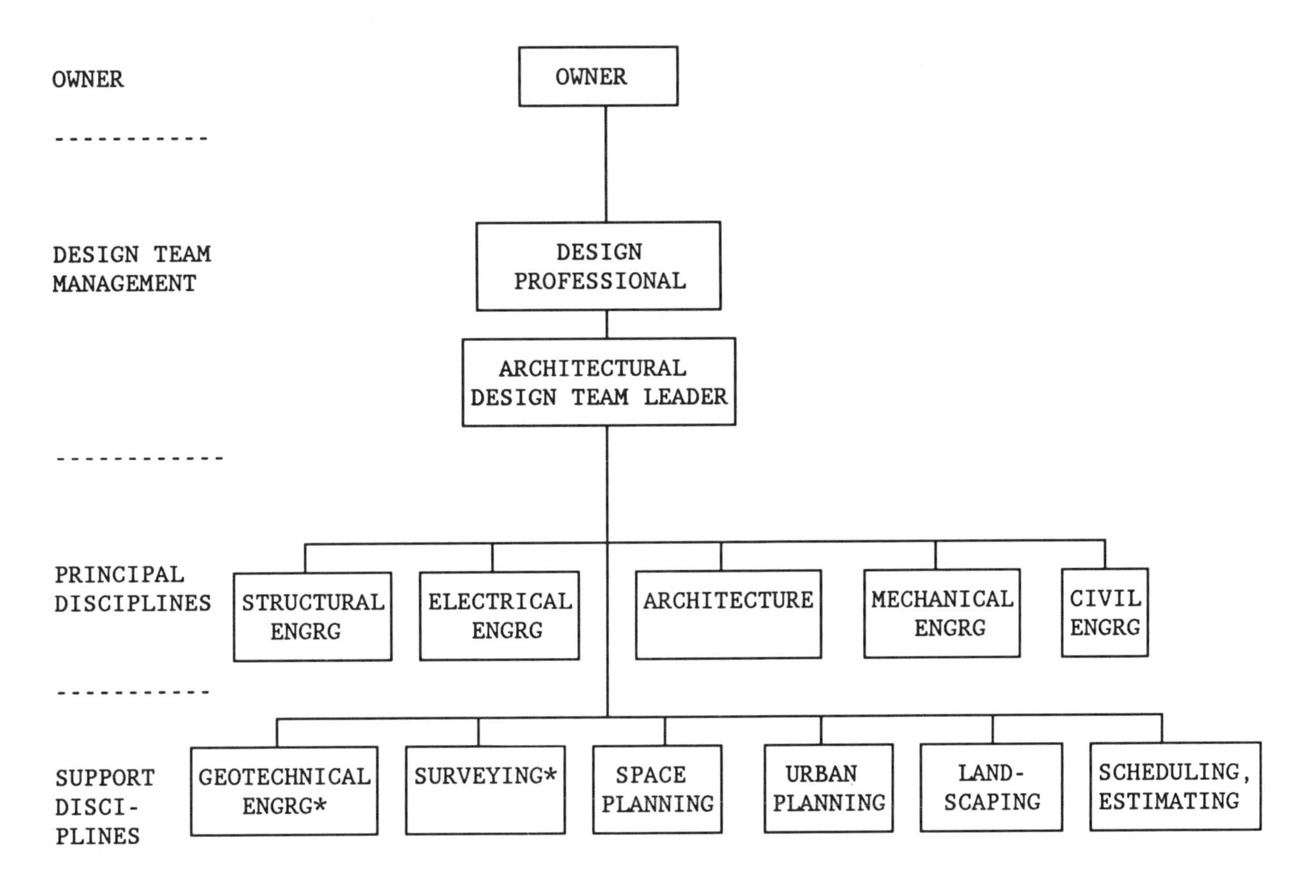

*=May contract directly with owner and supply information for use of design professional.

FIG. 9-2. Example of Multidiscipline Project Organization: Architectural Project

the project's functional requirements, consistent with budget, schedule, and other constraints. The project plan describes the overall requirements of the constructed project, including the owner's needs, site constraints, functional requirements, and budget. This document, combined with design codes and criteria, forms the project requirements.

The challenge for practitioners of the design disciplines is to develop a coordinated design that satisfies the project's requirements and is consistent with their own technical requirements. A coordinated design, however, may involve some compromises. For example, initial cost or operating efficiencies may be modified to accommodate the project requirements or the needs of other disciplines. The extent to which compromise is allowed, to accommodate any designer's disciplinary requirements, is dictated by the project requirements, except when safety is affected. A primary consideration of the team members from every design discipline is safety of the user, the site workers, the public, and the environment.

Other design considerations or priorities include the project requirements, functional requirements, compatibility with other disciplines, esthetics, initial construction costs, and operating costs. Attention to these considerations by the practitioners of each discipline and coordination of their efforts with those of the team members from other disciplines lead to design quality.

The need for coordination among the practitioners of different design disciplines and with the owner cannot be overemphasized. A coordinated design has the best chance of satisfying the project requirements with minimum conflict among the design elements. Furthermore, a coordinated design may reduce construction cost by requiring fewer field changes, may function more efficiently, and may be more esthetically

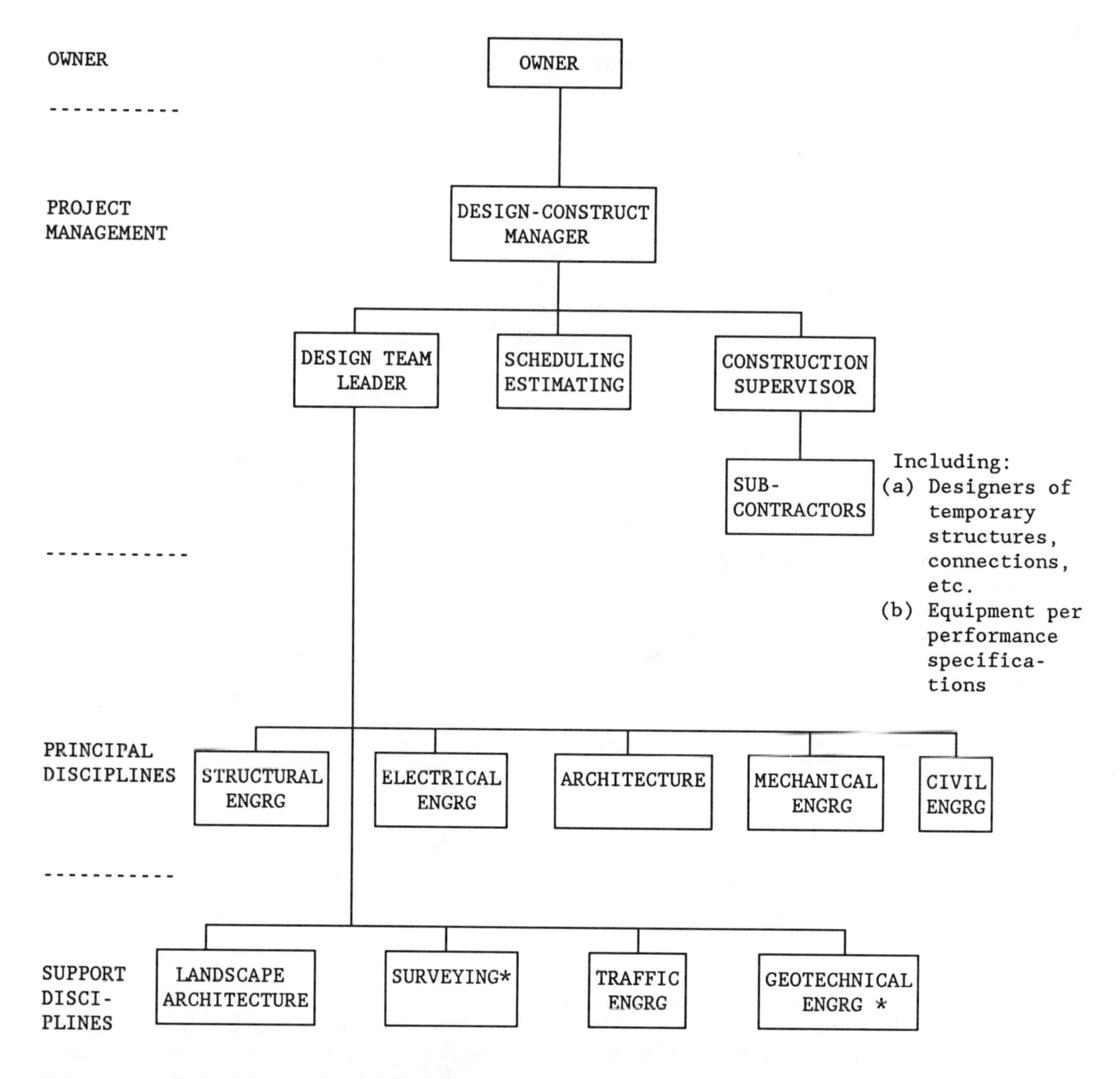

FIG. 9-3. Example of Multidiscipline Project Organization: Design-Construct Project

pleasing. The team members from each discipline are committed to meeting their own technical requirements without compromising those of the team members from other disciplines. In many cases, peripheral interests of each design discipline must be sacrificed for the overall interests of the project. While this may be easy to establish in principle, it is often difficult to execute during the active design effort. The realistic owner is aware of the need for compromises and participates actively in making difficult choices.

Contractual relationships among the owner and practitioners of the various design disciplines take many forms. Depending on the desires of the owner or the design professional, each discipline may have a separate subcontract, or, in the case of a large multidiscipline firm, discipline expertise may be supplied under the owner/design professional agreement without subcontracts. Sometimes there are direct contractual relationships between the owner and the various disciplines, although there must always be one design entity in charge to coordinate and provide technical initiatives. Contractual relationships define legal responsibilities and roles between parties. The basic project organization of disciplines and support disciplines will generally remain in effect. Each has a definite function in the design effort and each must do its part to meet the overall project requirements.

9.3 PARTICIPATION BY PROFESSIONAL DISCIPLINE LEADER DURING DESIGN

The professional discipline leaders are those individuals who lead their respective discipline's design work for a project. They are responsible for implementing the design services and reporting to the project design professional or design team leader. Technical decisions, staffing to meet technical and schedule requirements, supervision of the discipline team, and coordination with team members from other disciplines are all parts of the discipline leader's responsibility.

The professional discipline leader's responsibilities also include: (1) Adherence to contract obligations and project requirements; (2) technical correctness; (3) management of resources; and (4) conformance to codes and regulations.

The primary reason for a multidisciplinary effort is the array of specialized technical skills required to complete many project designs. Successful projects result when all disciplines are coordinated to achieve an integrated design solution that is consistent with project requirements. Each discipline professional, by virtue of design specialty, is responsible for correct technical applications within his or her discipline. These applications require proper use of engineering principles in the design solution as well as applicability and compatibility of the design to the overall project requirements. These requirements, which include constructability, maintainability, cost (both initial and operating), and esthetics, are considered and weighed critically to help achieve a quality design.

As a decision-making member of the design team, each professional discipline leader has a responsibility to assist other members in developing an integrated design solution. The leader considers not only the project's requirements applicable to a specific discipline, but also those applicable to other disciplines. Understanding of and commitment to the total project by all key participants are prerequisites to design quality.

Design quality is measured by: (1) Soundness of technical approach; (2) correctness of the numerical and graphic representations; (3) coordinated integration of the requirements of all disciplines into the constructed project; and (4) adequacy of specifications.

9.4 PARTICIPATION BY DESIGN PROFESSIONALS DURING CONSTRUCTION

The design professional's participation during the design phase of the project is defined in the contract with the owner. Quality in constructed projects is enhanced by a full-service design contract in which the design team's responsibilities continue after completion of design. Under a full-service contract, the design professional is responsible to the owner for the execution of the design team's contractual obligations during construction. A design professional's participation during construction is recommended, because design team members are the most knowledgeable about the intent of the design with respect to satisfying the requirements of each discipline as well as project requirements. Technical aspects of these services are the responsibility of professionals within each discipline.

The discipline leader's responsibility during construction is in several categories, as related to his or her particular discipline:

- Coordinate with other discipline leaders.
- Monitor and control discipline budget.
- Review submittals required by the construction documents.
- Evaluate alternative materials and designs.
- Evaluate modifications or changes.
- Visit construction site.
- Document reviews, site visits, and other activity.
- Perform final review and report results.

Review of constructor shop drawing submittals for conformance with the construction documents and the design objectives is an important part of the discipline leader's responsibility.

CONCLUSION

Most projects for design professionals involve more than one professional discipline. The goal of each multidiscipline design team is to provide a functional facility that meets the project requirements.

The design team for any multidiscipline project consists of practitioners of principal and support disciplines. The design professional has overall responsibility to the owner for the design, with individuals from other disciplines providing technical expertise and support to fulfill this responsibility. Each discipline, singly and cooperatively, is responsible for preparing a design that adheres to the project's requirements. The goal of the design team is the successful integration of all disciplinary technical requirements within the framework of safety and project requirements for a quality constructed project.

Key members of the design team are the lead practitioners of each discipline. They are responsible to the design professional for providing the services, adhering to the project's requirements, technical accuracy, quality in the design, and managing resources to meet the schedule and budget. They are also jointly responsible to coordinate their services with those provided by the team members from other disciplines.

CHAPTER 10
DESIGN PRACTICES

INTRODUCTION

Design practices involve office operations, the design professional's relationship with owners and constructors, design requirements, and programs for quality. Among other activities are those related to codes and standards, regulatory agencies, and grant applications.

10.1 OFFICE OPERATION

A successful design office's greatest resource is well-qualified personnel. Education and experience are the ingredients for these qualifications as well as for professional registration by state governments. Also, participation in continuing education programs and professional groups helps professionals stay up-to-date on current aspects of design practice, such as materials, design and analysis methods, computer systems for design and drafting, problems that have been encountered by others, and current business practices. In some cases, particularly for small offices, it may not be prudent to maintain a staff with experience in all practice areas encountered, and outside consultants may be needed.

The organizational structure of a design office is best developed in response to its particular needs and objectives, varying from time to time with the size of the firm, work being performed, and laws under which it must operate. Large offices find that an organizational chart assists the staff in understanding assignments, responsibilities, and authorities. Some general considerations of office operations are discussed next.

10.1.1 General Management of Design Office

General management includes hiring and retaining a qualified staff, procuring design contracts, arranging finances and accounting procedures, establishing and communicating goals and objectives, and establishing standard operating procedures.

10.1.2 Organization for Projects

One way to organize for management and execution of a project is to assign qualified personnel from each discipline to the project design team, under the guidance of the design team leader (discussed in Chapter 8).

Other firms prefer to organize on a departmental basis, a discipline basis, or a combination. Each approach has its strengths and weaknesses when applied to specific firms and projects.

10.1.3 Office Procedures

Establishment of policies and procedures, including office quality control procedures, promotes effective and efficient functioning of a design office, helps to provide a common base for all projects, and improves concentration on the owner's requirements for design and production of the contract documents.

The best use of standards supports rather than inhibits creativity. When the size, uniqueness, or cost of the project warrants and the owner and design professional concur, normal office quality control procedures may be augmented to provide greater control of the project.

10.1.4 Filing and Storing Documents

A system of filing and storing written material, which includes standard designations and classifications indexed for identification and location, increases office efficiency. Examples of documents to be filed include material regarding the development of the project scope and requirements, correspondence and reports, documentation of conversations or phone calls, design calculations, drawings and specifications, quality control forms, cost estimates, contract documents, schedules, time sheets, project costs, value engineering or life-cycle cost information, shop drawings and manufacturer's submittals, construction reports, and submittals to the owner or regulatory agencies. Large or complex projects may require separate files for each of the categories mentioned, while a single file is usually sufficient for small projects. Files may also be arranged by phases of projects.

Retention time for documents after completion of a project depends primarily on the likelihood of reference needs for future work or subsequent claims. Discarding duplications before filing, and reproducing data on disks or film will help to minimize warehouse storage or space requirements. Finally, a system of dating of the files facilitates periodic discarding of material no longer needed. Applicable statutes of limitations and repose enter into decisions to retain or discard files.

10.1.5 Reference Library

Current reference material, including costs, standards, manufacturer's catalogs, and design manuals, should be available to the design professional. Documents from previous projects can also be valuable. A system of cataloging this material for easy access will make a library more useful.

10.1.6 Drafting Practices

The design professional's drawings and specifications provide the constructor necessary information on the design concept, size and scope of the job, materials, performance, and quality requirements, as well as numbers and sizes of materials or items, and how they are to be assembled into a final project.

The preparation of drawings is referred to as "drafting," which includes computer-aided drafting, or CAD. The following are some items to be considered in the drafting procedure, many of which are performed with some degree of automation if CAD procedures are employed:

- Preparation of a schedule of planned drawings, outlining the content of each drawing.
- Preparation of rough layouts before beginning the drawings, to determine placements on the sheets and scales to be used. (Interdisciplinary coordination is easier if corresponding plans are drawn to the same scale and arranged similarly.)
- Use of precise and legible lettering, and allowing for possible use of reduced-size reproductions of the drawings.
- Use of standard symbols and abbreviations, and uniform terminology to enhance ease of reading and interpretation of the drawings.
- Use of clear, legible, and consistent dimensioning, and avoidance of repetitious dimensions, to help avoid conflict in the drawings.
- Coordination of the drawings with the technical specifications.

Many owners have developed their own drafting standards which they prefer, and discussing these standards with the owner before drawing begins prevents costly rework. If the design professional is using a CAD system of drawing production, modifying to owner standards may be time-consuming, and the owner may be willing to accept some of the design professional's standards to expedite the project.

10.2 RELATIONSHIPS WITH OWNER AND CONSTRUCTOR

Mutual respect, a close working relationship, and clear understanding between owner and design professional contribute to quality in the constructed project. Chapter 6 describes the Agreement for Professional Services between the owner and design professional, including necessary or desirable provisions of an equitable agreement. Provisions included in the agreement for a particular project should be as specific as possible, including, for example, scheduling or milestone dates, lists of services and materials to be provided by the owner, and lists of instruments of services to be provided by the design professional to the owner.

During the construction phase the design professional may represent the owner in certain matters, as defined by their agreement. Experience on many projects has clearly shown that quality is enhanced when designers can interpret their design documents in the field and are available to detect and adjust their designs to accommodate conditions different from the design assumptions.

10.3 DESIGN ACTIVITIES AND REQUIREMENTS

Quality design is based on engineering principles and professional judgment. In addition, a quality design meets the project requirements and satisfies applicable codes and standards (see Section 10.5). It is important that the owner fully understand the design services to be provided and be kept informed about the progress of the work through effective communication during the design process. Quality design consists of meeting the project requirements which often include, in addition to technical considerations, other factors relating to acceptance by users and the general public. Examples are security, appearance, generated noise, increase in traffic, and relationships of anticipated use to the surrounding neighborhood. Alternative studies (see Chapter 7) are often useful in finding economical solutions to societal needs in addition to those more narrowly associated with the project.

The design professional considers the requirements of the construction process (see 10.3.4—Constructability Reviews) and communicates the design intent to the constructor on the plans and specifications. A quality project includes communication and general agreement of term definitions among the owner, design professional, and constructor, as well as understanding and agreement on how to process change orders, clarify details, correct mistakes, and resolve conflicts. These considerations involve all parties to the project and extend throughout the construction phase of the project.

10.3.1 Design Considerations

Serviceability refers to those factors that affect the usefulness of a facility, including subjective perceptions of those who use the facility. The latter factors may or may not be covered by code requirements. Examples for buildings include vibration caused by mechanical equipment or by people walking, building sway due to wind, and transmission of sound to adjacent rooms or areas. Partial or total mitigation of these phenomena may increase the project cost, and the relative benefits and costs of alternative mitigation procedures should be discussed with the owner to obtain concurrence with procedures proposed for implementation.

Another important consideration for some owners is the life-cycle cost of the project. For example, an owner may prefer to pay greater initial costs for a mechanical system that will cost less to operate during the life of the project. Life-cycle cost determinations require analysis of initial costs, projected increases in costs, and the anticipated life span of materials, equipment and finishes. Often an owner will deliberately choose low first cost to lowest life-cycle cost; while the argument that choices should be made on the basis of life-cycle costs has many adherents, in the final analysis, quality is achieved when the design and construction match the balance selected and communicated by the owner.

An owner's desire to build quickly may result in a phased (or fast-track) approach to a project. This technique attempts to compress the overall project schedule by allowing construction on portions of the project to begin before design is completed for the entire project. Additional construction is then begun as design progresses. The possible advantages to the owner of this technique are lower financing and escalation costs (because of shorter design-construction period) and earlier use or occupancy of the facility. However, the owner must realize that design and construction contingency allowances need to be provided for unanticipated changes during design and construction and recognize that construction cost estimates will not have been based on completed plans and specifications.

10.3.2 Design Reviews

Many owners require reviews at various stages of the project (e.g., 10%, 30%, 60% complete) to determine if the project is being developed according to their requirements. Similar reviews by the project team, even if the owner has no such requirement, may save the design professional considerable time and expense and contribute to a quality design. Reviews by any and all parties are undertaken responsibly because many decisions and much subsequent work will be based on implied as well as expressed approval.

10.3.3 Construction Costs

Cost of construction is an important factor in owner decisions regarding project feasibility and financial planning. The design professional, although unable to guarantee construction costs, can provide an estimate of probable construction cost for use by the owner.

Often, the estimates of probable construction cost can be developed by the design office's regular staff members, whose experience is usually derived from actual involvement in similar projects. If this experience is not available inhouse, it may be desirable for the design professional to hire appropriate personnel or to contract out for the estimating services. Owners may often elect to develop their own estimates of construction costs. Sometimes it is desirable to review the construction cost with an experienced constructor, including allocation of enough time to study the plans thoroughly and do a complete take-off of quantities. In any event, the knowledgeable owner realizes that all cost estimates except proposals that a constructor actually tenders for acceptance are approximate. Unless specifically represented as a firm proposal, all "estimates" are to be regarded only as reasonable approximations.

Information on representative unit costs may be obtained from:

- Construction-cost indices appearing in authoritative industry publications.
- Local cost records published by various construction groups.
- Data on unit prices published by state agencies from time to time for various projects. These data are often a helpful cost gauge because they include local labor rates of various crafts, their availability in specific areas, work conditions, and material costs.
- Information from bids for similar projects. This type of information is especially valuable if it is available for specialty projects.
- Publications of the American Association of Cost Engineers.

10.3.4 Constructability Reviews

Constructability of a project refers not only to the adequacy of information on plans and in specifications to construct the project, but also to other aspects that can affect the work, such as site restrictions, economics of the proposed construction, availability of materials, construction equipment requirements, local work force available, and environmental considerations. Ability to construct includes practicability of achieving specified tolerances, access needed to make proper connections during assembly, and resolution of conflicts between architectural, mechanical, electrical, and structural requirements or elements. Often the issues involve maintenance of operations, utilities, etc. Three-dimensional models, both full and reduced scale, are sometimes used effectively as a method to study constructability, train and educate construction personnel, coordinate system interfaces, and eliminate interferences.

Continuing review by the designer for constructability aspects during the design phase may reduce the problems encountered during construction and possibly result in lower project cost.

10.3.5 Peer Reviews

Peer reviews are reviews of office or project procedures conducted by independent reviewers in an effort to enhance the quality of projects. Reviews of office operations and organization are normally performed by independent reviewers and may be beneficial in establishing an efficient, effective office operation.

Project peer reviews are concerned with a particular project and may be performed by an independent person or team not connected with the project. Project peer reviews are more common for projects involving innovative, unusual or unique systems and for projects greatly affecting the general public or environment. The owner may contract and pay for an independent peer review in a manner similar to arrangements for other project consultants. Sufficient time is needed in the project schedule to allow for the peer review. Peer reviews are further discussed in Chapter 21.

10.4 DESIGN-RELATED QUALITY CONTROL

Quality control requirements and structure for the project are discussed in Chapter 19. These procedures applied to design include checks, reviews, approvals, and work audits and are generally covered by standard office procedures. If the project is sufficiently large or complex a project-specific quality control program and manual may be justified. The goal of design-related quality control is quality in the design, rather than the generation of paper which attests that procedures or manuals have been followed. However, large projects can require more formality and a "paper trail" to document proper procedures.

10.5 COMPLIANCE WITH CODES AND STANDARDS

Codes and standards are developed by governmental units and industry or professional-technical associations to protect the public's health and safety. Rules are normally issued by a governing body to clarify and augment these codes and standards. Early identification of appropriate codes and standards can prevent reworking plans and specifications and save considerable cost and delay. Similarly, some stability in rules issued by governing bodies is necessary to allow owners, designers, and constructors to meet the requirements of agencies without hardship.

Since codes and standards typically address particular aspects of design and construction, the design professional can expect to find several codes and standards applicable to a project, including those of interest to civil engineers, electrical engineers, mechanical engineers, structural engineers, and architects. They also are normally related—and sometimes unique—to the regulatory agencies that have jurisdiction over the project. Various regulatory agencies and their roles in a construction project are discussed later in this chapter.

Applying codes and standards to design activities is sometimes difficult, especially for design professionals working on a project in an unfamiliar geographical area. Local and regional codes are common and are usually modified versions of national model codes; therefore, design based solely on a national model code may not always satisfy the requirements of a local authority. Design professionals can facilitate approvals and permits by obtaining the latest versions of the applicable codes and standards and by inquiring if there are any proposed revisions anticipated to take effect while the project is being designed and constructed.

Review for compliance with codes and standards by a knowledgeable discipline representative during the design phase but before submittal, will help to meet owner and regulatory requirements and develop confidence in the design team.

10.6 APPROVALS AND PERMITS BY REGULATORY AGENCIES

It is the design professional's responsibility to meet accepted professional standards, although approvals and permits issued by regulatory agencies are favorable indications of design quality. Preparation of plans and specifications which comply with codes and standards also will reduce time and effort in obtaining required approvals and permits.

Assignments and roles should be written, to reduce the possibility of misunderstandings. The agreement for professional services generally gives the responsibility for obtaining permits to the owner acting with the assistance of the design professional. Early identification of the appropriate local, regional, state, or federal regulatory agencies and their permit requirements will save time and effort, since different agencies have different levels of review and difficulties of compliance. If a variance from regulations is warranted, the appeal procedure described by the codes may be used.

Meetings with agencies to discuss their requirements and how these requirements may affect the design will be in the best interest of the project. Information from these meetings can then be used to determine the type of submittals required and the need for any sequential approvals by different agencies, and will facilitate establishment of a schedule for submittals and approvals. Agency review time can significantly affect an overall schedule, particularly the construction start date, and it is best to obtain approvals, when feasible, prior to the owner obtaining prices for the construction. This will reduce the potential for change orders resulting from regulatory agency requirements.

Copies of approval and permit applications, as well as copies of approvals and permits themselves, should be furnished to the other team members as appropriate. Sometimes copies of approvals or permits are required to be maintained at the construction site. Copies are needed by the design professional to provide final documentation of the work effort and to confirm that no additional requirements have been imposed upon the project by regulatory agencies. If additional requirements have been imposed, they can then be addressed by the design professional and handled by change order, if necessary.

10.7 GRANT PROCEDURES

Government grants may be necessary or desirable to develop a project. However, the owner and design professional need to be aware that grant procedures and requirements may significantly increase the time and efforts required of both parties and may restrict the project in quantitative and qualitative ways. Also, meeting special requirements to obtain a grant may increase construction costs, perhaps more than the value of the grant. Discussions with the owner prior to finalizing the agreement for professional services will help to determine the effort required of the design team in meeting the requirements of grant procedures.

10.8 DESIGN RESPONSIBILITY

Under state licensing laws, the design professional who seals the plans and specifications is responsible for the design. The designer, as a member of a design organization, commits the organization to this same responsibility. It is therefore important that the owner and design professional have a complete understanding of the extent, degree, and limits of services to be performed. This is usually accomplished in the written agreement for the design services.

When developing the scope of services, it is important for the design professional to have control over the design throughout the project, including the construction phase. If possible, such words as "complete design services" are to be avoided, since they are vague and subject to various interpretations. Use of standard agreements and contracts is much preferable since they are clearly understood. When asked to use nonstandard agreements or contracts, the design professional should obtain a legal opinion before the document is signed. The design professional should not sign documents or drawings that indicate a certification of the work or project without checking on legal responsibility and liability. Such precautions are in the owner's interest also, since an improper agreement can lead to misunderstandings, lack of communication, loss of respect, and poor quality.

The authority and responsibility of the design professional during construction should be defined as precisely as possible. For example, safety on the job site, adequacy of form work, shoring, and similar items are usually the responsibility of the constructor and are beyond the control and responsibility of the design professional. These responsibilities are set forth in greater detail in Chapter 16. The authority and responsibility of the design professional while on the construction site to interpret plans and specifications, clarify details, correct errors, and handle change orders are essential parts of a quality project and should be provided for in the agreement.

Federal, state, and local laws and regulations should be taken into account in agreements and contracts defining responsibility. State courts, legislatures, and registration boards are continually redefining responsibilities and legal liabilities of the members of the project team.

CONCLUSION

Design practice involves office operation, the design professional's relationship with owners and constructors, design requirements, and programs for quality. Also included are activities related to codes and standards, regulatory agencies, and grants. Establishing efficient office practices, observing quality standards, and closely following owner requirements and regulatory procedures are essential parts of a comprehensive program to achieve design quality in the constructed project.

CHAPTER 11

PRE-CONTRACT PLANNING FOR CONSTRUCTION

INTRODUCTION

This chapter deals with the owner's planning process in preparation for construction site activities. The key word is planning *for* construction, not planning construction operations. This planning activity takes place during all project phases prior to award of the construction contract and is interrelated with activities previously discussed, such as alternative studies, project team formation, project design, and preparation of construction contract documents.

The analysis and planning leading to the owner's decision on preferred contractual arrangements with design professionals and constructors include consideration of resources available for construction, regulatory agency requirements, time constraints, site selection and development problems, and other factors. The owner's capability to engage in and satisfactorily complete this process is vital to the project's success.

11.1 OWNER'S CAPABILITIES

In the simplest of projects the owner, as an individual, may do the project planning, perform the design, and construct the project. As projects increase in complexity the owner finds that assistance is required at each phase of project development, from setting project requirements to completion of construction. This assistance may be obtained by employing additional experienced individuals. A more practical approach to the problem is to contract with design professionals, construction advisers, and constructors as the project develops. These experienced advisers will guide the owner in making investigations and reaching decisions in planning for the construction of the project. The owner has the responsibility of seeking advice necessary to provide supplementary capability to make informed decisions on matters relating to planning for construction, including financial, regulatory, and legal matters.

11.2 RESOURCES FOR QUALITY CONSTRUCTION

Resources available to the owner, design professional, and constructor for the construction of the project may place constraints on project activities and influence decisions as to project requirements, planning and design, contracting strategies, and construction operations.

11.2.1 Financial Resources

The owner is responsible for providing funds to plan, design, and construct the project. The availability of total funds and the cash flow available will influence the project and are considered during the formulation phase. Some owners find it particularly difficult to provide sufficient funding for "up-front" activities such as preliminary planning, geological studies, subsurface exploration, alternative investigations, and other activities leading to the establishment of project requirements and design criteria. Adequate funding of these efforts results in a well defined project and reduces the risk of "surprises" during design and construction due to unforeseen conditions. Such surprises are often looked upon as lack of quality in design and construction, when they are actually due to inadequately defined and communicated requirements.

Each member of the project team has an interest in the financial health of the other members, especially as to any matters which may influence performance during design and construction of the project. Consideration of financial capabilities impacts the structure of owner/design professional and owner/constructor contracts.

11.2.2 Materials for Construction

The availability and cost of materials for construction influence planning, design, and construction operations. The following questions need to be considered regarding natural materials at or near the site:

- What type of foundations best suit existing soil and geologic conditions?
- Can these conditions be modified to accept other types of foundations?
- How would flooding or erosion affect

materials used and construction operations?

- Is aggregate for concrete available on the local market? Must it be manufactured; or imported?
- Are fill materials available on site? What are other alternatives?
- How do site locations (i.e. congested or remote) influence choice of materials and methods of construction?

Consideration of manufactured materials will pose other questions:

- Are the materials readily available at a competitive cost? Shortage of steel? Shortage of cement? Lead times for critical components?
- How do transportation costs at a remote site influence material selection?
- How do storage and preparation of materials at a congested site influence construction operations?
- Can unique or specially prepared materials be provided on a realistic schedule for installation by local constructors?
- Will local preference or "buy American" legislation influence selection of material and equipment?

The choice of materials for construction is necessary early in the planning and design process and requires input from all members of the project team, including the owner. If the constructor is not yet selected, appropriately qualified advisers on construction operations are consulted by the owner.

11.2.3 Manufacturing Capability of Suppliers

Specialized permanent equipment, components, and materials may require sophisticated or specialized manufacturing capabilities to meet the project requirements. It may be beneficial for the owner, assisted by the design professional and other advisers, to review and evaluate manufacturing capabilities of suppliers of specialized items.

Steps to be taken in assessing manufacturing and delivery capabilities are:

- Review project requirements to identify demands for specialized manufacturing capability.
- Survey potential suppliers to evaluate capability to meet demands.
- Review quality control programs of manufacturers.
- Consider the possibility of independent observation, expediting, and inspection of manufacturing and testing.
- Plan to include explicit requirements for the needed capability in the procurement documents.
- Plan to accept bids or offers from only those firms prequalified to meet manufacturing capability requirements.

11.2.4 Human Resources

In the initial stages of project planning the owner, assisted by advisers, may evaluate the human resource needs of the project. The continuity of key professional and management personnel and the availability of skilled work force are factors contributing to the quality of the project.

Continuity of the design professional and constructor personnel through all phases of the project, including construction, start-up, and operation, generally enhances communication and project understanding among all team members. Consultation with project design team members will assist in the interpretation of plans and may prevent the issuance of change orders contrary to the design intent.

During the initial stages of planning for project construction the owner, assisted by advisers, can evaluate work force availability at the site. Discussions with local construction firms, labor unions, vocational training facilities, building contractor associations, and others help to determine availability of a skilled work force needed to construct the project. Steps to mitigate the effects of projected work force shortages may include:

- Discuss action to be taken with labor unions or trade associations.
- Modify project facilities (e.g., worker housing at remote sites).
- Plan for additional construction supervision personnel.
- Plan for additional quality control measures.
- Initiate plans for a training program.
- Plan to design project modules to be factory fabricated and assembled for transport to the site.

11.3 REGULATORY AGENCY REQUIREMENTS

The impacts of regulatory agency requirements on planning for construction are considered throughout the planning and design phases. Areas of agency input which affect project planning include site and job safety, minority employment requirements, schedules to meet agency requirements for grants and loans, use and disposal of hazardous materials, and protection of public health and safety. Nonroutine requirements set by an agency call for early evaluation by the owner and the design team. If regulations seriously impact the project as envisaged, alternative approaches may be formulated and discussed with agency staff for preliminary approval. The vagaries of the code review process and the subjective nature of codes pose additional risks of which the owner needs to be aware. Unless otherwise agreed, arrangements for obtaining required permits in a timely fashion are the owner's responsibility.

11.4 SITE DEVELOPMENT

Site development activities which may take place during the planning and design phases but before completion of the construction contract documents include:

- Construction of access roads and rail sidings.
- Extension of utilities to the site.
- Construction of independent utility systems, if required.
- Construction of temporary buildings and other facilities.
- Construction of laydown areas and fabrication yards.
- Planning for traffic control.
- Construction of detour routes.
- Relocation of utilities, highways, rail, and other facilities.

Road and utility construction contracts may be issued separately from and prior to the contract for project construction; utility extensions and relocations may be performed by the utility involved; or construction required for site development may be included in the project construction contract. Consideration of scheduling, site congestion, construction sequencing, cost, and other items will influence decisions as to contractual arrangements.

11.5 REVIEW OF DESIGN-CONSTRUCTION ALTERNATIVES

At one or more appropriate points in the evaluation of planned alternatives (Chapter 7), alternatives are reviewed from the standpoint of construction sequencing, plant/equipment layouts, constructability, safety considerations, and construction work plan development. During the design phase of the selected alternative, similar reviews in greater depth are conducted, designs are modified where appropriate and practical, and special problems are addressed in drafting the contract documents. The owner is continually involved in these matters in order to remain informed about the project and refine requirements as the options are identified.

11.6 CONTRACTUAL ARRANGEMENTS

The owner, as the prime mover in the project, establishes the contracting strategy. The strategy is planned to meet the project requirements and is formed to reflect the owner's capabilities and the required capabilities of the design professional(s) and constructor(s). Elements considered are the project size, the capabilities and sophistication of the local construction industry, availability of work force, legal and political requirements, owner's time constraints, and financing. The selection of the correct constructor(s) and design professional(s) and use of the right contract terms is important. Quality depends on the competence and integrity of each team member.

The owner has the option to structure contracts for the project in many different arrangements:

- In the traditional arrangement, the owner/design professional contract, and the owner/constructor(s) contract(s) provide for the design professional to design the facility and to prepare the contract documents, including plans and specifications, to be used by the owner in contracting with the constructor to construct the facility.
- In the design-build arrangement, the owner contracts with a design-build firm to perform both design and construction of the project. Turnkey construction, an extension of the design-build arrangement, may include provision for the constructor to provide land acquisition, financing, purchase and installation of industrial process equipment, and other services not included in a standard design-build contract.
- In the design-construct approach to a project, the owner and staff plan, de-

sign, and construct the project. The staff may or may not be reinforced and supplemented by professional service firms or labor contractors to supply personnel to work under the owner's direction for design and construction of the contract.

- Any combination of these arrangements which fuses the capabilities of the owner, design professional, and constructor may result in quality in the constructed project.
- Privatization is an option being used by public owners which combines many of these contracting methods.

Payment terms for any of the contracts written by the owner with other project team members may be on the basis of lump sum or reimbursable cost plus fee and may or may not include performance incentive clauses. For specific discussion of structuring agreements for design services and for construction, see Chapters 6 and 14, respectively.

CONCLUSION

Planning for construction, which takes place prior to the completion of plans and specifications, is primarily concerned with possible constraints on project construction which may influence the owner's decisions on matters relating to project requirements, schedule, and budget. Evaluation of such factors as the owner's own capabilities, resources required for construction in the way of financial capability of team members, material and equipment availability, and human resources guide the owner in determining contractual arrangements with advisers, design professionals, and constructors for project planning, design, and construction.

Contractual arrangements available to the owner include the traditional arrangement of the owner/design professional agreement coupled with owner/constructor contract(s), as well as design-build and owner/constructor arrangements, together with combinations of such. Arrangements for construction of site utilities and access may be handled by a set of separate contracts to be performed before design is completed and before project construction at the site begins.

During planning and design of the project the impact of design options on construction sequencing, plant/equipment layouts, constructability, and safety considerations should be reviewed.

CHAPTER 12

THE CONSTRUCTION TEAM

INTRODUCTION

The goal of the team structured for the construction phase of the project is to build a facility with quality, safely, on schedule, and within budget. Like other phases of the project, construction requires a team effort of skilled individuals and qualified organizations working toward the common goal.

Assembling a construction team structured within the framework of the project team (Chapter 3) for field operations at the construction site(s) is the owner's responsibility. The composition of the construction team is determined by project requirements, capabilities of the owner and each of the other team members, and by contractual arrangements (Chapter 14).

In addition to key personnel representing the owner, design professional and constructor, staff members of regulatory agencies will have input at the construction site. Other contributors include suppliers, subcontractors, and craftsmen, as part of the constructor's organization. Advisers on financial, insurance, and legal matters may assist in the construction efforts of the team as well as the activities of individual team members. Each participant (organization or individual) is held responsible for the work he or she does. Teamwork is necessary, with each member carrying out assigned responsibilities in a quality way.

12.1 CONTRACTUAL ARRANGEMENT

In planning for construction (Chapter 11) the owner, assisted by advisers, sets the contracting strategy for the project. The contracting arrangement chosen will influence the composition of the construction team.

Contracts, by their nature, attempt to define and control diverse interests of the parties involved. When possible, contracts should seek to align interests and emphasize mutual goals. To this end each contract is structured to define clearly scope of work, roles and responsibilities among parties, coordination and communication requirements, division of authority, and other contractual relationships.

The owner is responsible for:

- Developing written requirements.
- Developing clear and equitable contracts.
- Assembling qualified team members.
- Providing finances, land, rights of way, etc.
- Paying invoices promptly.
- Signing contract change orders promptly.
- Enforcing contract terms and conditions.

Although a variety of contracting arrangements can be used, field operations at the construction site conform to a general organizational pattern shown in Figure 12-1.

12.2 FIELD ORGANIZATION FOR CONSTRUCTION

The organization and interrelationship of construction team members is shown in Figure 12-1. Note that the position of the boxes shows project team members in the traditional contracting arrangement, but the arrangement may also apply to internal organization in the case of a design-build constructor type, or an owner's staff performing design and construction.

12.2.1 Owner's Resident Project Representative

A key member of the construction team is the owner's resident project representative (RPR) who may be the owner, the project manager (Chapter 3), a member of the owner's staff, a member of the design professional's staff (resident engineer), a member of the constructor's staff (design-build arrangement) or an independent construction management firm employed by the owner. The source of the RPR is determined by the contractual arrangements chosen by the owner, and the RPR's authority and responsibility are defined by contracts executed by the owner with other team members.

The authority and responsibility of the own-

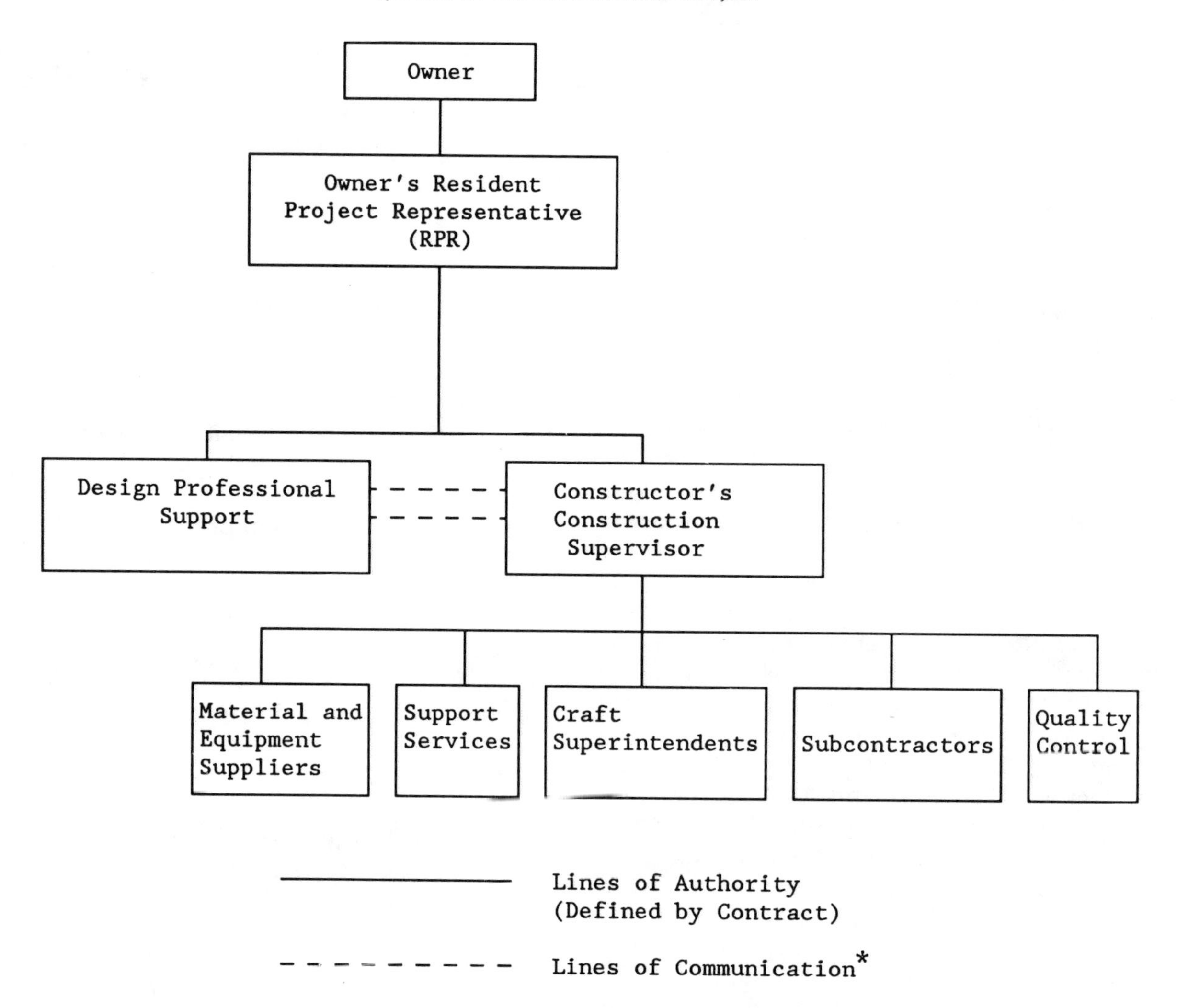

FIG 12-1. Construction Team: Field Organization Chart for Traditional Project
This figure shows a general organization pattern for field operations at the construction site.

er's RPR are defined specifically in the construction contract documents and generally limited to site activities. In this role the RPR:

- Represents the owner at the construction site.
- Mediates conflicts at the construction site.
- Administers the owner's contracts with the constructor, design professional, and others.
- Reviews and evaluates deviations from existing schedules.
- Reviews and approves invoices and forwards them to the owner for payment.
- Reviews and approves change orders to construction contracts and forwards them to the owner for signature.
- Supervises other personnel in the employ of the owner engaging in activities at the construction site.
- Communicates appropriately with regulatory agencies to the extent required by law or contract.
- Conducts periodic progress meetings.

12.2.2 Constructor's Construction Supervisor

The constructor's key employee on the site is the construction supervisor. Depending on the size and complexity of the project, this position may be filled by the constructor or a member of the constructor's staff. The construction supervisor, acting within the constraints of the contract with the owner:

- Represents the constructor at the site.
- Provides submittals required under the contract.
- Manages execution of the work.
- Arranges for licensed professionals to design such temporary facilities as are not included in the designer's design (e.g., shoring, scaffolding, coffer dams, river diversions, traffic maintenance, etc.), and makes appropriate submittals to owner and regulatory agencies.
- Manages subcontractors, suppliers, and artisans.
- Arranges for timely payment of constructor's employees, subcontractors, and suppliers.
- Manages the constructor's quality control program.
- Manages the job-site safety program.
- Coordinates and complies with regulatory agency requirements to the extent required by law or contract.

12.2.3 Design Professional's Construction Support Services

The design professional's staff functions in the design professional's office and in the field to provide support services specified in the owner/design professional contract. The design professional does not have a contractual relationship with the constructor, hence has no formal means of communication with the constructor except through the owner's RPR, or except as where otherwise provided in the construction contract documents. Exception to this is generally made on technical matters relating to the constructor's contract submittals or routine exchanges of information. The RPR should be kept informed of the content of such direct communications. The design professional's contract with the owner generally delegates authority and responsibility to:

- Resolve technical questions on clarification of plans and specifications.
- Review contract change orders from a technical viewpoint.
- Review technical data (including shop drawings) and comment appropriately.
- Evaluate alternatives and substitutions proposed and make recommendations to the RPR.
- Coordinate with regulatory agencies on technical questions.
- Provide observation of construction when required by contract with the owner or by law.
- Review plans for temporary structures required for construction only as to impact on the completed project.

12.2.4 Regulatory Agency Participation

During the course of construction, representatives from agencies of federal, state, and local government may visit or be assigned to the construction site. These observers and inspectors are responsible for determining compliance with codes and laws, rules and regulations of the appropriate agency. The objective of this activity is protection of public health and safety and of the environment.

Categories of inspection involving agency employees at the site are:

- Compliance with building codes and standards.
- Compliance with health and safety regulations.
- Compliance with regulations relating to disposal of hazardous materials.
- Compliance with regulations regarding air and water pollution.
- Certification and documentation of compliance with regulatory requirements.

Other agency personnel may be interested in visiting the project to observe progress as it relates to expenditure of public funds under grant programs, nuisance abatement under court orders, potential operational problems affecting public health and safety, and other matters of public interest. Prompt compliance with agency authority and coordination and communication with agency personnel by both the RPR and the constructor's construction supervisor will ease problems and contribute to a quality project.

12.2.5 Advisers for Construction

In the conduct of business the owner, design professional, and constructor separately retain advisers on matters of law, insurance, public relations, and other specialties. It may be appropriate to provide access to such advisers to assist the RPR on matters affecting the project as a whole during the construction phase. Immediate attention to problems which may later result in constructor claims, third-party litigation, citations by regulatory agencies, labor disputes, and public dissatisfaction will be cost-effective, speed contract completion, and help in producing a quality facility.

12.3 ASSEMBLING CONSTRUCTION TEAM

The owner is responsible for assembling the construction team. The process starts with the owner's decision to construct a project, continues through planning, design, and preparation of construction contract documents, and ends with execution of the owner/constructor contract and selection of the RPR.

Unless the project is very simple or the owner's staff is well qualified, the owner cannot proceed without competent advice on design and construction activities. In the traditional arrangement the owner first engages the design professional to assist in defining the project requirements and to perform project planning and preliminary design. Analysis of planning and preliminary design as it relates to construction operations is done by the constructor if available or by special advisers or qualified members of the design professional's staff.

After the project has been defined, the owner is in a position to select the preferred contractual arrangements outlined in Chapter 11 and to authorize the design professional to complete project design and prepare plans, specifications, and other contract documents for project construction. In the traditional system, the owner invites constructors to submit proposals or sealed bids for project construction. On private projects, a select list of prequalified bidders may be invited. Although many governmental agencies are required by law to engage in open bidding and award procedures, it is possible in many cases to structure a prequalification procedure which requires constructors to demonstrate capability by record of experience, previous performance, financing, and availability of experienced management and skilled supervisory personnel and necessary construction equipment.

The owner's RPR is required to be in place when the construction contract is awarded. It is preferable, however, for the owner to make this selection early in the project planning and design process so that the RPR is knowledgeable not only of decisions made but of why they were made. As an experienced member of the team, the RPR can contribute actively to the planning and design phases of the project. The RPR's qualifications include:

- Experience in project construction.
- Training with design professionals and constructors.
- Knowledge of the owner's organization.
- Ability to communicate project requirements, keep participants informed, gain commitment from construction team members, and respond to their needs.
- Recognition of and respect for legitimate interests of each participant.
- Dedication to quality.

CONCLUSION

The owner is responsible for selecting a construction team capable of completing a quality facility on time and within budget while protecting safety of the public, site personnel, and the environment. The RPR selected by the owner is key to providing team leadership as he or she administers the contracts with the constructor, design professional, and others in the interests of the project.

Different contractual arrangements are available to the owner depending on type and size of the project and capability of the owner. In any arrangement, contract documents define relationships between the owner and other parties and delegate responsibility and authority to team members. Wherever possible, contracts are structured to align interests of the parties and are administered to foster teamwork in meeting common goals of producing a quality project.

Quality depends on the competence and integrity of each team member.

CHAPTER 13

PROCEDURE FOR SELECTING THE CONSTRUCTOR

INTRODUCTION

The selection of a qualified constructor to execute the construction at a competitive cost is an owner's decision that significantly influences quality in the constructed project. In reaching this decision, the owner formulates selection procedures and determines the constructor's ability to produce the desired results. In some cases this may include setting prequalification criteria for constructors. The owner is generally assisted in this effort by the design professional, legal counsel, and other advisers.

Many competitive bidding procedures are presented in documents prepared by professional organizations, industry associations, and by federal, state, and local agencies. These procedures, recommendations, laws, and regulations are most highly structured for competitive bidding of public work. Using public work procedures as a base, other procedures, such as competitively priced proposals and competitive bidding by a select list of bidders are also discussed in this chapter.

Constructors are motivated to bid projects (public and private) if the rules are fair, clearly stated, and followed.

13.1 SELECTION PROCEDURES AND QUALIFICATIONS

The procedure used by the owner in selecting the constructor may be informal or highly structured, depending on the size and complexity of the project and on the type of owner involved. Private-sector owners may award construction contracts on the basis of prices offered informally, even orally, by constructors known to them. On the other hand, most public agencies are required by law to use competitive bidding procedures. The owner, often assisted by the design professional and other advisers, formulates the procedure to be used to attract and evaluate interested constructors.

The first step in the selection procedure, either public or private, may be an examination of the qualifications of a number of constructors, invited by public notice or private means to submit qualifications to perform the project construction. The tailoring of the qualification appraisal will vary, depending upon the other steps in the selection process, i.e., competitive bidding, reimbursable cost-plus-fee contract, negotiated lump-sum, or unit-price contract.

For competitive bidding under mandated regulations of some public agencies, the only indicators of responsibility required of constructors submitting bids are generally: (1) Valid state license to perform the work; and (2) financial capability to enter into contract as evidenced by the ability to supply bid and performance bonds. However, public agency regulations may permit a more restrictive prequalification if the project requires specialized construction techniques, is on a critical schedule, or has special characteristics regarding protection of the environment and safety of work.

A general checklist of information which may be requested by the owner for examination is:

- Constructor name, location, license, corporate structure, if applicable.
- Business data—financial information, bonding capacity, bank references.
- Construction experience—projects constructed or under construction—size, type, performance on schedule, and budget.
- Work force on constructor payroll vs. subcontracting.
- Equipment owned vs. rentals.
- Safety records—written program.
- Quality control—written program.
- Resumes of key executive and supervisory staff.
- Current work load—remaining bonding capacity.
- Personnel to be assigned to key positions of management and field supervision.
- Performance of completed projects—references.
- Record of litigation.

Associated General Contractors of America (AGC) has prepared a preprinted form, "Construction Contractor's Qualification Statement for Engineered Construction" (AGC Document #220), for use as a suggested generic prequalification statement or a contract-specific qualification statement.

In examining and evaluating the information submitted, the owner may, in addition to checking the references supplied by the constructor, make independent checks on the constructor's credit standing, visit projects completed and interview owners and operators, investigate safety and litigation records, and review other elements of performance.

13.2 CONSTRUCTOR QUALIFICATION

As a result of the submittal and confirmation of constructors' qualifications the owner is in a position to proceed with selection of the constructor. Private owners may complete the process if they choose to employ the constructor or design-build constructor considered to be best qualified, or they may use the information to develop a "short list" of constructors to:

- Bid competitively for the work on a unit-price or lump-sum basis.
- Present priced proposals for negotiation of a contract on mutually acceptable terms.

If the owner is a public agency, the analysis of the qualifications information may be used to select a short list for submittal of priced proposals. If prequalification information is not being requested from constructors, criteria may be set by the agency for qualifying bidders in terms of experience, size, licensing, work load, specialized expertise, financial status, and other qualifications appropriate for the construction of the project.

13.3 SELECTION BY COMPETITIVE BIDDING

Two ingredients for a true price competition bidding are: (1) A clear, concise set of construction plans, specifications, and other contract documents; and (2) a defined procedure for the bidding and award process. The contract documents specify the product to be delivered by the constructor. The bidding procedure protects the owner as well as the bidders and preserves the integrity of the bidding process in a step-by-step approach to bidding and awarding the contract.

13.3.1 Role of Design Professional in Competitive Bidding

The degree of involvement of the design professional in the bidding process depends upon the owner/design professional agreement for services. The terms of this agreement are influenced by the owner's staff size and capability. In a typical arrangement, the design professional:

- Prepares plans, specifications, and other contract documents as well as bidding documents for the review and approval of the owner, the owner's legal counsel, and other advisers.
- Assists the owner in obtaining bids or priced proposals for each separate contract to be awarded.
- Assists the owner in evaluating bids or proposals and in awarding contracts.

The responsibilities of performing activities and preparing documents for the bidding process can be listed and assigned in the manner illustrated as part of Appendix 3.

13.3.2 Competitive Bidding for Public Work

Competitive bidding is used as a method of constructor selection for federal, state, and local projects. For these types of projects, the use of competitive bidding is usually mandated by law or regulations. This mandate reflects a legislative body's opinion that competitive bidding for construction contracts provides value to the taxpayers and fairness to the construction industry when spending major sums for public contracts. The goals of value and fairness are met when the integrity of the bidding process is preserved.

Beyond the general mandate to use competitive bidding, there are numerous specific rules and criteria that vary with the public owner involved. The information required for typical bidding and award of a public construction contract is defined by the set of documents issued to prospective bidders. These documents generally include:

- Bidding documents which apply to the bidding process, including: the invitation to bid (legal notice), instructions to bidders, information for bidders, and bid forms. Bidder qualification data may be a part of the bid submittal if proof of valid licensing and bonding capacity is all that is required. If constructor prequalification is used as a means of developing a list of qualified bidders, this activity occurs before the start of the conventional bidding procedure.

- Contract documents which specify the constructor's performance on the project and generally include: the owner-constructor agreement, performance and payment bonds, the bid, general conditions, supplementary conditions, specifications, drawings, and any addenda issued during the bidding period.

Discussion and guidelines relating to the content and structure of the documents listed, as well as other aspects of the bidding and awarding process are covered by publications of EJCDC, AIA, AGC, CSI, and other professional organizations. (Refer to Appendix 2.) Regulations and laws applying to federal (FAR), state, and local agencies also prescribe the content of contract documents and bidding procedures.

13.3.3 Bidding Procedures for Public Work

The procedures required to solicit and inform bidders, to receive and analyze bids, and to award contracts under a competitive bidding system include the following sequential actions to be taken by the owner, assisted by the design professional after consultation with legal counsel and other advisers.

13.3.3.1 Prior to Bid Opening

- Receive and evaluate constructor qualifications if a prequalification of bidders is part of the process.
- Invite qualified constructors to bid on the project through legal notices and other advertisements, direct mailings, and notice to trade publications and accredited plan rooms.
- Set bid-opening time and date, to allow sufficient time for constructors to make quantity takeoffs, investigate the site, receive subcontractor prices, determine material and equipment costs, and take what other actions are necessary to prepare a unit price or lump-sum bid.
- Arrange for distribution of bidding and contract documents to interested bidders, accredited plan rooms, and other viewing locations.
- Maintain a current list of document holders.
- Make appropriate arrangements so that prospective bidders may have access to the site.
- When appropriate and practical, hold a pre-bid conference at the site to answer inquiries on and clarify provisions of the bidding documents. The pre-bid conference is not used to convey information in addition to that contained in the bidding documents. The constructor and owner are held responsible for what is in the written documents, not the conversation at the pre-bid conference. If clarification is required, an addendum is issued to all document holders.
- Issue addenda to all document holders of record. If an addendum is required, and cannot with certainty reach all prospective bidders in time to permit adjustments in the bid to be submitted, the bid opening may be postponed.

13.3.3.2 Bid opening

- Require that all bids be dated and time stamped when received at the bid-opening location. Late bids should be returned unopened.
- Open bids at a public meeting where they are read aloud.
- Check bid submittals for presence and amount of bid security, acknowledgment of receipt of all addenda, presence of required documentation.
- Make original bidding documents available for inspection in the presence of the bid-opening official.
- Safeguard bids for later evaluation.

13.3.3.3 After Bid Opening

- Prepare bid tabulations and make information available to interested parties.
- Verify and analyze qualification data submitted with the bids.
- Confirm compliance with other requirements of bidding documents.
- Take appropriate action with advice of legal counsel in rejecting non-conforming bids.

- Take appropriate action with advice of legal counsel in permitting withdrawal of bids by bidders claiming errors in bid preparation.
- Analyze price bids, supporting information, and documentation, using criteria set forth in bidding instructions, and determine lowest responsive, responsible bidder.
- Reject all bids if constraints of budget, schedule, or other critical elements cannot be met.

13.3.3.4 Contract Award

- When required, obtain approval of federal or state agencies administering grants and/or loans.
- Make award within the time specified in the bid documents (AGC recommends a maximum of 75 days after the letting.) If the time must be extended, a written approval from the low bidder is required.
- Prepare a notice of award, forwarding multiple copies of the contract documents for the successful bidder's signature. The notice of award allows a certain period of time for the constructor to execute the documents and return them along with executed bonds, certificates of insurance, and other required documentation.
- Place owner's signature on the executed contracts, and issue notice to proceed with the work.

Careful observance of procedures, rules, and regulations protects the integrity of the bidding process, provides fair and equal treatment of all bidders, and gives the public agency owner fair prices from the competitive process.

13.3.4 Competitive Bidding for Private Work

The private sector owner may elect to follow essentially the same procedures in securing competitive bids as has been outlined for the public sector. The owner may invite a broad spectrum of the construction industry to participate, or may limit participation to a select bidders list chosen by prequalification.

In either case, the owner and bidders rely on defined bidding and contract documents, generally prepared by the design professional, to provide mutual understanding of the project and to set rules and procedures for competitive bidding and award of contract.

13.4 COMPETITIVE SELECTION PROCEDURES FOR NEGOTIATED CONTRACTS

Situations occur where structured price bidding may not be appropriate. In these situations competition takes place first in a comparison of qualifications submitted by interested constructors. The owner may select the constructor best qualified for the assignment and proceed through the negotiation of a contract for project construction. This approach generally results in some form of reimbursable cost-plus-fee contract, although negotiated lump-sum or unit-price arrangements are not precluded.

If competition more directly related to construction or design-build services is desired, the owner selects a list of constructors on the basis of qualification analysis. The owner then solicits proposals from constructors on the select list on the basis of certain proposed elements defined by the owner. These proposals may include some of the following:

- Understanding of the project.
- Approach to project—work plan—unique ideas.
- Organization of project activities—services proposed.
- Proposed schedule, with milestones.
- Programs for safety, quality control, design and use of temporary structures.
- Availability of crafts, use of subcontractors, minority involvement.
- Use of local resources.
- Business information—wage and salary costs, overhead costs, contracting policy, insurance, etc.
- Project budgets by components of the work.
- Proposed cost of work—unit-price, lump-sum, reimbursable cost-plus-fee.
- Key management and supervisory personnel to be assigned to the project.

With this information available the owner makes an evaluation of the organizational and cost elements of the work and negotiates a contract with the constructor judged to have the best overall proposal.

13.5 NONCOMPETITIVE SELECTION FOR NEGOTIATED CONTRACTS

Under certain circumstances the owner may select a specific constructor and negotiate the contract. Situations where this procedure applies include:

- The owner may choose the constructor on the basis of the constructor's satisfactory performance on work previously done for the owner.
- The constructor may have unique qualifications for the work to be accomplished.
- The urgency of the situation—damage control, restoration of failed utilities, protection against flood or other natural disaster, or other circumstances—requires immediate action.

Good public policy dictates that, as a general principle, government entities should use competitive bidding procedures except where circumstances dictate otherwise, such as in cases of extreme emergency or specialized procurements.

In the case of sole source award of a contract by a public agency, the constructor has unique characteristics such as a record of successful business relations with the owner, specialized expertise not available elsewhere, or immediate availability in case of emergency.

CONCLUSION

Quality in the constructed project is achieved when the constructor conscientiously follows the plans and specifications in completing the construction contract safely, on time, and within budget. With this requirement in mind, the owner formulates a selection procedure which places major emphasis on obtaining and evaluating the qualifications of the constructors available for the project. The owner may use this information to prepare a list of prequalified bidders or proposers eligible to compete in the constructor-selection procedure.

The role of the design professional in the selection of the constructor lies in preparing, for the owner's approval, the bidding package, including contract documents which define the project and the procedures for submitting price-competitive bids or proposals. The design professional then assists the owner in administering the bidding process, in evaluating the bids or proposals received, and in awarding contracts.

Complex bidding procedures are used by public agencies concerned with obtaining a competitively low price from a responsible bidder under a fair and impartial bidding process. Less complex procedures are often used by private-sector owners soliciting bids from a preselected list of constructors. The competitive bidding process usually results in the contract's being awarded to the qualified bidder with the lowest unit-price or lump-sum bid.

Constructors may also be selected on the basis of competitively priced proposals, as well as by a sole-source selection process based on continuing business relationships with the owner, unique capabilities, or immediate availability.

AGC is committed to competitive bidding practices and has worked with EJCDC in development of a "Construction Contractor's Qualification Statement for Engineered Construction"—a standard qualification statement, and "Recommended Competitive Bidding Procedures for Construction Projects" (see Appendix 3).

CHAPTER 14

CONSTRUCTION CONTRACT

INTRODUCTION

The construction contract documents—the contract with attachments—define the agreement made between the owner and the constructor for project construction. This is a two-party agreement which does not include the design professional, although the design professional acting under the owner's contract for professional services may be authorized to perform certain of the owner's responsibilities under the construction contract.

Care is taken that the contract, in whatever form, accurately documents a "meeting of minds" of the parties, is stated in clear, concise language, clearly defines responsibilities of the parties without overlaps or voids, and aims at achieving a quality project.

14.1 CONSTRUCTION CONTRACT DOCUMENTS

The term "contract documents" includes the contract for construction between the owner and constructor, together with other documents referenced by and made a part of the owner/constructor agreement. These documents, taken together, define the responsibilities of each party during the construction phase. Contract documents generally include:

- Contract.
- General conditions.
- Supplementary conditions.
- Plans and specifications.
- Addenda (if any) issued before bid submittal.
- Constructor's bid.
- Notice of award.
- Performance and payment bonds.
- Change orders or contract modifications issued.

The design professional, under contract with the owner, is generally responsible for preparing and assembling the contract documents for review and approval by the owner and legal counsel.

14.2 CONTENT OF CONSTRUCTION CONTRACT

Although an extensive checklist and discussion of items to be included in the construction contract is beyond the scope of this guide, the following are key issues that should generally be included:

- Definition of the parties.
- Identification of applicable law.
- Effective date of the contract.
- Definition of uncommon terms.
- Scope of work.
- Plans and specifications for facilities to be constructed.
- Contractual milestone dates/completion dates.
- Assignment of authority and responsibility between parties.
- Risks and liabilities assumed by each party.
- Terms and methods of payment.
- Insurance and bonds.
- Contract change-order procedure.
- Settlement of disputes.
- Contract termination conditions.
- Indemnifications.
- Warranties/guarantees.

14.3 FORM OF CONSTRUCTION CONTRACT

The contract documents usually include four basic elements: (1) Bidding documents; (2) contract forms; (3) contract conditions; and (4) plans and specifications. The contract conditions normally include the use of a standard general conditions set, together with supplementary conditions intended to modify the standard general conditions to meet the specific requirements of the project. The specifications generally include all technical requirements for the work.

The question of where to place or find a specific subject in the contract documents has been addressed in a joint publication of EJCDC and AIA titled "The Uniform Locations of Subject Matter and Information in Construction Documents." This document has been significant in developing a uniform approach to the organization of contract documents. In fact, this document and the complete coordination of the EJCDC and AIA forms in accordance therewith are compelling reasons why the EJCDC or AIA series should be used wherever possible.

14.4 STANDARDIZATION OF CONSTRUCTION CONTRACTS

Most large organizations, including public and private owners, design professionals, and constructors, have individually drafted standard contract forms for individual use. Professional organizations and industry associations such as EJCDC, AIA, AGC, APWA, and CMAA visualize an advantage to the construction industry in the broad use of standardized contract content, forms, definitions, and language. The EJCDC, AIA, and AGC particularly have spent a great deal of effort in developing and endorsing standard forms, contracts, general conditions, and other documents, together with commentaries on the use of this material (see Appendix 2). Some of these documents have been printed in volume so that multiple sets of contract documents can be assembled using the preprinted copies.

14.5 INTERNATIONAL CONSTRUCTION CONTRACTS

A widely recognized standard form entitled "Conditions of Contract for Works of Civil Engineering Construction," has been prepared by the International Federation of Consulting Engineers (FIDIC) in consultation with lending institutions and with constructor associations. This form, recently revised and reissued in its Fourth Edition (the "Red Book"), is considered to have fairly balanced contractual risks and responsibilities between owner and constructor. It is frequently the design professional's responsibility to adapt its Part II, the Conditions of Particular Application, to meet the needs of particular countries or projects. A valuable guide to the use of FIDIC conditions of contract was published in 1989 and is available through the American Consulting Engineers Council.

CONCLUSION

The owner/constructor contract, together with its attachments, defines the project and reduces to writing the agreement between the parties for construction of the project. Constructors are required to carry out their work in accordance with the contract documents. Clear, concise plans, specifications, and contract language are necessary to describe a "meeting of minds" between the parties.

The design professional is not a party to the owner/constructor contract but is generally responsible under the owner/design professional contract for the preparation of the construction contract and its companion documents. Before release for bidding or contracting purposes, these documents are reviewed and approved by the owner with advice of legal counsel.

Substantial efforts by professional and industry associations have advanced the cause of standardization of contract content, form, and language with a large body of information. Coordination of all legal documents as to responsibilities, relationships, definitions, schedules, and contingency measures is critical.

CHAPTER 15

PLANNING AND MANAGING CONSTRUCTION ACTIVITIES

INTRODUCTION

There are many ways to plan and manage project construction activities. The traditional project team members—owner, design professional, and constructor—may vary their roles depending on the option selected.

Primary sources of authority regarding allocation of responsibilities are the owner/design professional agreement and the owner/constructor contract, provided these documents reflect allocation of responsibilities under statutory and code law.

Essential elements of planning and managing construction include clear communications through planned reporting, scheduled meetings, memos, shop drawing processing, and review of progress payment requests. Project management tools include formulating and regularly updating the construction plan and schedule, estimates, and the quality control program.

15.1 ORGANIZATION FOR PROJECT CONSTRUCTION

The project owner's list of responsibilities as they affect the construction phase of the project includes:

- Arrange funding for the project.
- Select the contracting arrangement.
- Define the role of the design professional in the construction phase.
- Select the constructor and enter into the construction contract.
- Obtain necessary approvals from regulatory agencies.
- Administer contracts and coordinate activities of involved parties.
- Make prompt decisions on construction matters.
- Respond promptly to materials submitted for the owner's review.
- Make prompt payment under terms of contracts.
- Enforce contracts.
- Other duties and responsibilities assigned by contract.

Responsibilities of the design professional in the construction phase, in addition to responsibilities under the law, are determined by the owner/design professional agreement. They generally include:

- Interpret and clarify contract documents when questions arise.
- Review and approve technical elements of contract change orders.
- Review and comment on technical elements of contract submittals (as specified in the agreement for professional services).
- Provide advice to the RPR on technical elements of design and construction.
- Consult with the RPR on the level of quality control required on the project.

If the limits of responsibility and authority are clearly defined and provided for under the owner/design professional agreement, the design professional or members of the design professional's staff may:

- Function as the owner's RPR.
- Review and approve change orders for the owner's signature.
- Review and approve progress pay estimates for payment by the owner.
- Review and take appropriate action on the constructor's submittals required by the contract.
- Observe progress and quality of constructed work.
- Represent the owner with regulatory agencies.
- Provide other professional services specified by the contract.

The constructor is responsible for:

- Site safety.
- Means and methods of construction.
- Construction sequencing and scheduling.
- Quality control related to construction activities.

- Management of and payment to his or her suppliers and subcontractors.
- Construction of the facility in accordance with contract plans, specifications, and approved change orders.
- Compliance with other sections of the owner/constructor contract.
- Project construction schedule and budget.

15.2 PRECONSTRUCTION MEETINGS

After the construction contract has been signed, a preconstruction meeting hosted by the owner is recommended. The meeting includes key construction team representatives from the owner, design professional, and constructor. Representatives of principal subcontractors and regulatory agencies may be invited to attend.

A meeting agenda should be prepared, and accurate minutes kept and distributed. Typical agenda items for a preconstruction meeting are: introductions; lines of communication and submittals, including correspondence; site rules and regulations; safety and first aid; procedures for issuing and revising design information and authorizing changes; survey information; constructor's designated areas and coordinating procedures; methods and schedules of payment; certificates of insurance; procedures for overtime and shift work; security; cleanup; temporary facilities and services; project-schedule program; project-cost program; material handling; the Equal Employment Opportunity (EEO) Act; environmental procedures; the owner's role and responsibilities; specific state and local laws or regulations; claims, disputes procedures; subcontractor approval; community relations; critical specifications status; quality control; submittal procedures, and other subjects.

After the owner's preconstruction meeting, the constructor may host a similar meeting attended by representatives of subcontractors, material suppliers, vendors, and others in his or her support group. The RPR and the design professional may be invited to attend, but only as observers or sources of information for the constructor. Meeting procedure and agenda items are similar to the owner's meeting, but are discussed with particular emphasis on performance of the support group effort and its commitment to the project through the constructor.

Both preconstruction meetings are structured to develop common goals and lines of communication for the different participants involved in the project, i.e., team-building.

15.3 CONSTRUCTION ACTIVITIES

Early construction planning permits schedule milestone and interface dates to be included in the construction contracts. Furthermore, owner-supplied facilities, services, and programs can be defined. Once construction begins, plans are carried to a greater level of detail and revised as necessary. This refinement is accomplished by the constructor with review by the RPR.

15.3.1 Construction Schedule

During the construction phase, the schedule is refined and expanded to include subcontractor's schedules and the work activities. For long construction projects, rolling schedules are used, with periodic development of detailed schedules covering a period of several months. Daily schedules are needed for direction of the work force. If multiple construction contracts are used, each constructor's schedule is integrated with the overall schedule when the constructor mobilizes. The type, style, and level of detail are specified by the owner.

Computers can be used effectively to integrate levels of detail and individual schedules into one composite schedule. Effective scheduling software packages simplify this task. Turnaround time is minimal and new graphic capabilities make the schedules easier to understand. Integrated schedules may be analyzed and a critical path determined.

Significant delivery dates and issue dates of design elements can be added to the construction schedule to integrate design and construction. By including construction submittals and equipment deliveries as work items in the schedule, major suppliers are included in the project schedule. On a typical project containing mechanical and electrical equipment, much of the project is actually constructed in factories far removed from the site, and follow-up on manufacturing and delivery are critical for completion. The schedule identifies interface points between constructors and subcontractors, and significant milestones, including change orders. It can also account for scheduling of utility hookups and changeovers, and the transfer of loads and functions from temporary to permanent facilities. Schedule durations are based on known quantities and realistic production rates. Work force loadings are determined to verify the validity of the schedules. The schedule provides for the input of change-order work and an analysis of its impact or restraint on other activities. Schedules are expressed as bar graphs/Gantt charts for simple projects, while Network Analysis Systems (NAS) such as Critical Path Method (CPM) or Precedence Diagramming Method (PDM) are recommended for more complex projects.

15.3.2 Estimates and Cost Control

During the construction phase, the constructor's estimates should be refined and updated by the constructor. Work productivity is determined to identify problem areas. Identifying and, if necessary, correcting cost trends early in the process prevents overruns and disputes.

With lump sum contracts, a procedure for payment based on value of work accomplished as per contract is determined prior to starting work. Unit prices may apply for some work, and for other work predetermined percentages are indicated.

15.3.3 Construction Facilities and Services

Prior to mobilization, plans are finalized for construction facilities and utility services. Owner-supplied facilities should have been identified and designed prior to the constructor bidding or selection phase. The task during the construction period is to install these facilities and utility services as soon as possible and assign operation and maintenance to the constructor.

15.3.4 Material Management

The constructor develops a plan for the project that determines the amount of purchasing, receiving, special storage, and in-storage maintenance of materials required, and the time frame. Results of this analysis permit sizing the offices, warehouses, and laydown yards, as well as determining the staffing required. A system may be initiated that will key material deliveries to the project schedule or, if a computerized material management system is utilized, assign schedule item numbers to the material records. The availability of materials is a prime consideration any time the schedule is revised. Productivity and quality often suffer when crews are started and stopped repeatedly due to material shortages or for any other reasons.

15.3.5 Work Force Management

From schedule of effort data the constructor determines the number of people required by craft and special skills and determines the availability of work force to meet needs. If project needs appear to exceed the available work force, the constructor considers means of expanding the work force pool. Various incentives are available, including higher wages, relocation expenses, per diem living expenses, and overtime schedules. Training programs may be the most dependable solution, if time permits. A revision of the subcontracting plan which permits elements of the work to be performed off-site may also be considered.

15.3.6 Safety and First Aid

The constructor is normally responsible for planning and execution of a safety and first-aid program. The program is structured to comply with federal, state, and local laws and regulations.

Job-site plans may include:

- Posting safety rules, inspection procedures, and enforcement actions.
- Conducting safety training sessions.
- Training for first aid and fire fighting.
- Posting of emergency telephone numbers for paramedics, ambulances, fire fighters, and police.
- Definition of rallying areas (for head count) and delineation of escape routes in case of emergency.

15.3.7 Other Activities

Other planning and management activities may include site environmental control, hazardous waste handling, traffic control, EEO programs, and public relations.

15.4 COORDINATION AND COMMUNICATION

A successful construction phase requires coordination between design, construction, and start-up of the facility. The owner is responsible for providing this coordination in the contractual arrangement selected.

Communication is enhanced if lines of authority are followed. Some basics are:

- Only the constructor directs work.
- Only the constructor coordinates subcontractors.

- The constructor receives direction only from the owners, under terms of the construction contract.
- Manufacturers receive direction only from the party signing the purchase order, unless the contract assigns that authority to a second party. Transfer of authority is written into the purchase order.

CONCLUSION

There are many ways to plan and manage project construction activities. The traditional project team members of owner, design professional, and constructor may perform various roles depending on the option selected by the owner. However, certain responsibilities cannot be altered without threatening quality in the constructed project. These are:

- The owner is responsible for contract enforcement, and stopping work (except in emergencies).
- The design professional is responsible for design and design changes and interpretation of design requirements.
- The constructor is responsible for construction means, methods, sequences, direction of work, job safety, and completing the project construction to a level of quality in accordance with the requirements of the contract documents.

Essential elements of planning and managing construction include clear communications through planned reporting, scheduled meetings, correspondence, shop drawing processing, and review of progress payment requests. Project management tools include formulating and regularly updating the construction plan and schedule, estimates, and the quality control program.

Many other elements of construction activities require careful consideration, including site facilities and utility services, safety and first aid (including public safety), project cleanup, and public relations. Final closeout involves joint participation of the owner, design professional, and constructor.

CHAPTER 16

CONSTRUCTION CONTRACT SUBMITTALS

INTRODUCTION

After award of a construction contract, the constructor develops and submits certain information necessary for the construction process to the owner, or to the design professional for the purpose defined in the owner/design professional agreement. The required submittals are identified in the construction contract documents and may be classified generally as:

- Contract documentation.
- Initial technical documentation.
- Shop drawings for structural components.
- Shop drawings for manufactured structural or architectural components.
- Shop drawings for mechanical and electrical components (e.g., piping, electrical raceways, HVAC ducts, etc.).
- Shop drawings for equipment.
- Placing drawings for concrete reinforcing steel.
- Shop drawings for temporary construction.
- Results of independent testing.

This chapter describes planning, scheduling, preparation, and processing of construction contract submittals. It also discusses each type of submittal listed and suggests responsibilities and authorities for the various parties involved with submittals. In practice, the authority and responsibility of each party are defined by law (e.g., building codes, licensing laws, etc.) and the various project contracts; and nothing in this chapter is intended to alter responsibilities imposed by law and regulations, or to change or limit contractual provisions to which the parties may agree.

16.1 GENERAL

The professional services contract between the owner and design professional, who may also be the structural engineer of record, and the construction contract between the owner and constructor should define clearly the authorities and responsibilities of each party, including design services, scope and purpose of contract submittals review by the design professional, and scope of work to be performed by the constructor and subcontractors, such as fabricators, detailers, suppliers, and manufacturers, so as to avoid misunderstandings and vague, implied, or implicit responsibilities.

General procedures for processing various types of submittals are similar, but there are variations, as will be seen in the sections of this chapter that discuss each type of submittal. In particular, the procedure for shop drawings for structural components is more detailed because of the need to address design of connections. Also, the procedure for shop drawings depicting temporary construction is different from the other types, because the constructor has full control and responsibility for these shop drawings. The design professional is not normally involved in the design of temporary facilities or in the review and approval of shop drawings for them.

16.2 PLANNING SUBMITTALS PROCESS

Realistic planning by the owner recognizes that preparing, coordinating, reviewing, and approving shop drawings and other submittals is a necessary part of a project, and allows sufficient time and funding for the design professional and constructor and their subconsultants and subcontractors to accomplish their respective contractual obligations in this regard.

Before construction begins, the constructor and others having responsibilities for submittals should meet with the design professional to review design requirements, general procedures, lines of communication, and criteria for developing and processing the submittals.

Where the nature of the project requires it, the parties may agree that subcontractors and suppliers may communicate on technical matters directly with the design professional. Requests directed to the design professional for additional information or clarification of technical requirements should be in writing; but if unusual circumstances do not afford time for a written request, then oral communications are acknowledged in writing by the subcontractor or supplier and confirmed by the design professional. All other communications should be routed through the constructor. In any event, it is important that all affected parties be informed.

Time and effort spent by the constructor and the design professional in planning the submittals process will help to establish clearly delineated and efficient procedures and will greatly facilitate the preparation and processing of construction contract submittals.

16.3 SCHEDULING, PREPARING, AND PROCESSING OF SUBMITTALS

The constructor, assisted by fabricators and suppliers prepares a time schedule for submitting shop drawings or other submittals and, in cooperation with the owner and design professional, develops a schedule for reviews and approvals and a monitoring procedure.

Preparation of submittals is a responsibility of the constructor, but in the case of shop or placing drawings, preparation is usually actually performed by a detailer employed by or under subcontract to the constructor, fabricator, supplier, or manufacturer. Coordination of this effort is important, and in the event the detailing is subcontracted, the constructor or fabricator may find it advantageous to have a drafting supervisor in direct contact with the detailer to enhance communications.

The constructor (with advice from his or her fabricator, detailer, or supplier) may make a written request for approval if specific deviations from the construction contract requirements are necessary to facilitate material selection, fabrication, erection, or placement. The design professional reviews and recommends that the owner approve these deviations when appropriate.

If computer technology used in certain pre-engineered components, products, or stand-alone specialty items precludes the preparation of shop drawings, then the design professional and the constructor or supplier can review the technical requirements contained in the construction contract documents and determine whether a manufacturer's or supplier's certification of these items will satisfy the requirements.

A requirement that a structural steel fabricator, precast or pretensioned or prestressed concrete manufacturer, prefabricated components producer, or other product manufacturer be certified to perform its work may be included by the design professional in the construction contract documents. Certification, if required, would be in accordance with the specifications, a quality certification program of the American Institute of Steel Construction (AISC), Prestressed/Precast Concrete Institute (PCI), governmental agency, or applicable building code.

The design professional provides sufficient information in the construction contract documents to permit the preparation of shop or placing drawings. The drawings are then prepared by the detailer in accordance with design information in the construction contract documents, instructions from the design professional (or, in some cases for structural components, the fabricator's professional engineer), sound practice of the fabricating or reinforcing steel industry, and applicable regulatory requirements. The constructor reviews submittals prepared by his or her subcontractors or suppliers for compliance with contract documents, and certifies that the contents of each submittal conform with the contract documents unless clearly indicated otherwise.

The constructor then makes the submittals and the owner and/or design professional reviews the submittals and indicates acceptance or a need for revisions. Usually, the owner reviews items of a contractual or business nature, and the design professional reviews items of technical nature. Some items may be reviewed by both the owner and the design professional.

The design professional reviews submittals for conformance with the design concept of the project and information given in the construction contract documents, but does not review those aspects of a submittal that pertain to the construction process, such as the means, methods, techniques, sequences, and procedures of construction; detailing dimensions; fit or erectability in the field; or safety precautions and programs.

After review as provided by contract, the constructor returns the shop drawings, placing drawings, or submittals to the appropriate subcontractor.

16.4 CONTRACT DOCUMENTATION

Information or documentation required by the construction contract documents promptly after award of the contract may include:

- Performance, payment and material bonds.
- Proof of insurance coverage.

- Names of proposed subcontractors.
- Estimated cash flow requirements for the project.

Regulatory agencies may require additional documentation or information on such matters as project safety, wages and hours, minority employment, or environmental impacts.

16.5 INITIAL TECHNICAL DOCUMENTATION

Documentation of a technical nature needed early in the project by the design professional, or engineer of record, may include:

- Specifics of equipment and material to be incorporated into the project.
- Detailed schedule for performance of the work.
- Breakdown of any lump-sum bid items for partial payments.

16.6 SHOP DRAWINGS FOR STRUCTURAL COMPONENTS

Structural shop drawings depict components that will be part of a completed structure and are fabricated or constructed according to specific requirements as defined in the construction contract documents by the design professional, or engineer of record. This section is intended to cover a broad range of structural components and their connections, including steel, concrete, wood, and other materials. Examples are structural steel members, precast or pretensioned prestressed concrete structural members, light wood framing, and heavy-timber construction.

Detailed design of connections, or of unique structural elements, often is deferred until a fabricator is selected, so that the design can be customized to fit the particular capabilities and shop procedures of the selected fabricator.

The owner may contract with the engineer of record to design the entire completed structure, including the connections; or for structures with simple connections (shear connections only), to analyze and approve connections selected by the fabricator from accepted standards as indicated in the construction contract documents. The engineer of record would have authority and responsibility for the design, and the fabricator would have authority and responsibility for detailing the structural components and connections, but not for their design.

Alternatively, for more complex structures (structures with other than simple connections), if the engineer of record does not design all the connections, the construction contract documents must provide that the fabricator, as a part of his or her work, have a qualified professional engineer design or supervise the design of all connections not completely designed on the construction contract documents. The fabricator must provide certification from a professional engineer to that effect, to the engineer of record, who then reviews and approves the shop drawings. In this event, the construction contract documents include the necessary connection loading information as well as any performance data and loading requirements not defined in the codes and standards governing the project and which are needed to design the connections. Under this alternative arrangement, the engineer of record retains responsibility for design of the completed structure, for reviewing and approving the design of connections and structural components, and for review and approval of shop drawings, and the fabricator and his or her professional engineer have responsibility for their work.

In either alternative, design of connections is to be performed by or under the supervision of a qualified professional engineer, and the engineer of record is responsible for including sufficient information in the construction contract documents to permit the preparation of shop drawings. Review and approval of shop drawings, calculations, and associated documentation by the engineer of record shall be for conformance with the design concept of the project and information given in the construction contract documents and for the effects of the connection design on the primary structural system.

The constructor and his or her subcontractor, the fabricator, are always responsible for providing the materials specified and completing the requirements of the fabrication process—including maintenance of the specified fabrication and construction tolerances, development of detailed dimensions, and fit and erectability of the structure in the field—in a workmanlike manner and in accordance with the contract specifications, approved shop drawings, and standards.

16.7 SHOP DRAWINGS FOR MANUFACTURED STRUCTURAL COMPONENTS

Manufactured structural components that will be part of a completed structure are specified by the design professional, or engineer of record, in the construction contract documents. Examples are skylights, elevator structural supports, curtain walls, proprietary space-truss systems, steel stairs, precast concrete stairs, steel joists, light wood floor or roof trusses, cellular floors, decks, precast concrete components, and other pre-engineered components where design and fabrication are part of the scope of supply of the manufactured component.

The constructor has responsibility for complying with the requirements of the construction contract documents, and the manufacturer has responsibility for design of its product, and for supplying it in accordance with the required quality, including performance, material selection, and workmanship requirements.

For these manufactured items, the role of the design professional is to specify performance requirements and review the documentation provided by the manufacturer. The design professional may specify in the construction contract documents that the manufacturer's submittals bear the seal of a professional engineer. The design professional also compares the manufacturer's certifications of the item with the requirements of the construction contract documents and reviews the shop drawings, but only for compatibility with the completed structure.

16.8 SHOP DRAWINGS FOR MECHANICAL AND ELECTRICAL COMPONENTS

Many mechanical and electrical components, such as piping, electrical raceways, HVAC ducts and tanks, etc., are uniquely engineered for a particular project. Shop drawings for these components are considered in a manner similar to those specified for structural components by the particular engineering discipline involved.

16.9 SHOP DRAWINGS FOR EQUIPMENT

Design professionals may define the general nature and quality of certain manufactured or shop-fabricated equipment to be incorporated into the project by specifying a particular brand name and model (or its approved equivalent), or they may define the requirements by reference to trade industry codes or operational characteristics, such as operating efficiency, capacities, power requirements, or energy output. Examples of such equipment are pumps, electric motors, turbines, and heat exchangers.

Compliance with performance requirements contained in the construction contract documents may be demonstrated by a manufacturer's warranty and certification that the proposed equipment has been tested and complies with the requirements, or by the owner's representative witnessing tests in the manufacturer's laboratory and attesting to the validity of the data.

In reviewing shop drawings for factory-assembled materials and equipment, the design professional determines compliance with the design concept and compatibility with other elements of the project, such as:

- Anchor bolt layout.
- Foundation designs.
- Pipe fittings, flanges, and welds.
- Routing of utilities.
- Drainage for water and other liquids.
- Wiring for power supply.
- Interfacing of instrumentation and controls.

Although the design professional may review and comment on interface data, it is the constructor's responsibility to complete the installation in accordance with requirements of the contract documents.

16.10 PLACING DRAWINGS FOR CONCRETE REINFORCING STEEL

Placing drawings illustrate reinforcing steel components that will be part of a completed structure and are furnished and placed according to specific requirements as defined in the construction contract documents by the design professional, or engineer of record. Examples of components for which placing drawings are prepared are cast-in-place concrete and posttensioned prestressed concrete.

The design professional has authority and responsibility for overall design of the completed structure and for review and approval of the placing drawings for conformance with the design concept of the project and information given in the construction contract documents.

The constructor and his or her subcontractors have responsibility for preparing the placing drawings, providing the materials specified, and completing the requirements of the fabrication and construction process in a workmanlike manner and in accordance with the construction contract documents, approved placing drawings, and accepted standards.

16.11 SHOP DRAWINGS FOR TEMPORARY CONSTRUCTION

Temporary construction shop drawings depict components that will not be part of a completed structure, but are to be used temporarily during construction, such as sidewalk slabs, temporary lifts, temporary buildings, shoring, reshoring, form work, bracing, scaffolding, dewatering, or temporary power. The constructor has full authority and responsibility for these shop drawings, including design, preparation, review, and approval, since he or she develops the construction plan and has control of the construction process of which the temporary construction is a part. Construction or erection procedures, shoring, bracing for excavations, or other temporary construction requiring engineering analysis or design requires the seal of a qualified professional engineer affixed to the drawings and specifications.

The role of the design professional or engineer of record is to evaluate the effect of temporary structures on integrity of the completed structure. From this viewpoint, he or she may indicate in the construction contract documents how long such items as temporary bracing, shore-up, reshoring, and similar items need to remain in place. If a non-self-supporting frame is involved, and needs special treatment during construction, or if the design concept of the structure limits the construction sequence, he or she will also include this information. The design professional usually does not review temporary construction shop drawings except when necessary to determine compatibility with the design of the completed structure.

16.12 RESULTS OF INDEPENDENT TESTING

For some projects, the construction contract documents may require testing of certain materials by an independent testing laboratory to determine if they meet the requirements of the construction contract documents. Examples are soils testing, materials testing, chemical or biochemical testing of water, and shop inspection and testing of pipe fabrication.

The independent testing laboratory is employed by the owner or by the constructor with the owner's approval. The test results are submitted to the owner, design professional, and constructor simultaneously, and reviewed for adequacy by the constructor and for conformance with contract requirements by the design professional.

CONCLUSION

After award of contracts, the design is further detailed and defined as the constructor selects subcontractors, fabricators, materials and equipment suppliers. Detailing of structural connections and other elements is prepared once the fabricator is selected. Operating characteristics, foundations requirements, utility connections, and space necessary to provide for shop-manufactured and shop-assembled equipment units are determined and attested to by the equipment supplier. Materials testing is usually conducted and certified to by an independent laboratory.

The constructor is responsible for making all submittals required by the construction contract, aided by subcontractors, material and equipment suppliers, fabricators, and testing laboratories.

The owner is responsible for administering the contract and for reviewing submittals relating to contract terms such as insurance coverage, legal responsibility, schedules, change orders, extra work. Within the owner's organization the owner is responsible for developing a system for tracking submittals and for completing reviews on schedule.

The role of the design professional with respect to review of contractor submittals is specified in the design professional's agreement with the owner. Under terms of this agreement the design professional may be responsible for review of submittals relating to conformance with requirements of the technical specifications, and may also have responsibility in the review, change orders and extra work, and review of plans for temporary construction as they affect the design concept. The design professional is responsible for tracking submittals and completing reviews as scheduled. Other responsibilities include coordination of design disciplines on submittal review and interpretation of the intent of the contract documents in response to questions from the constructor.

The constructor is responsible for developing a system and schedule for preparing and tracking submittals required under the contract. The system must be realistic in evaluating requirements of the owner and design professional or engineer of record, as well as those of the constructor's subcontractors and suppliers. If regulatory agency approval is required for submittal reviews, this is accounted for in the schedule and work-flow plan developed by the constructor.

CHAPTER 17

CONTRACT ADMINISTRATION PROCEDURES FOR CONSTRUCTION

INTRODUCTION

The construction contract documents define the owner/constructor relationship and govern the performance of these parties during the construction phase of the project. The owner's resident project representative (RPR) is responsible for administering the contract. In this activity the RPR institutes procedures, monitors progress, and maintains appropriate records of the constructor's performance under and compliance with terms of the contract documents. The RPR is also responsible for meeting the owner's obligations under the contract, such as:

- Maintaining quality commitments.
- Timely payments to constructor.
- Timely review and comments on contract submittals.
- Timely decision on problems arising from unforeseen conditions.
- Coordinating site activities among owner's independent contractors.
- Planning, reviewing, and monitoring project communication, including progress reports, record-keeping and data retrieval systems.

Overall construction contract administration begins in the project's early stages and continues long after completion. Neither preconstruction nor postconstruction phases are included in this chapter. Furthermore, this chapter does not cover the management or administration of internal staffs, such as office and field design professionals and technicians, nor does it cover management or administration of subordinate service activities, such as testing laboratories, soils engineers, or surveying companies.

17.1 QUALITY COMMITMENTS

The RPR is responsible for implementing procedures for documenting review and evaluation of quality requirements specified in the contract documents. Construction quality generally involves two broad aspects: specified properties for materials, and workmanship. Materials can be further subdivided into in situ materials and procured materials.

17.1.1 Materials

In situ (natural or original) materials typically include native soils and rocks and often require laboratory testing and engineering evaluation of material properties to determine acceptability for project needs. Such laboratory reports and engineering evaluations become part of the project file. Retesting and other necessary follow-up analysis also become part of the file.

Procured materials are manufactured items, such as structural steel, asphalt, concrete, paint, glazing, mechanical and electrical equipment, and distribution systems, and can be evaluated and accepted by several considerations. The specifications outline the level of quality and the manner of qualification, if any, that will be required. For example, a manufactured product may be accepted based on the verification of brand name and catalog number, whereas a material such as paint primer may require a series of physical and chemical analyses to verify that specified requirements have been met. The procurement specifications state minimum standards.

Organizations such as the American Society for Testing and Materials (ASTM) and the American National Standards Institute (ANSI) publish extensive and reliable data on qualification and acceptance standards for numerous materials and products. Such standards clearly outline the specific qualification procedures required, include acceptance requirements and frequency of testing. It is the responsibility of the RPR to verify that the procurement qualification requirements are met.

Each procured item of material or product should be represented by a file listing the qualification procedure and minimum requirements and include the type of tests performed, the date the test was performed, the signature of the person performing the test, test results, any nonconformance reports and, if required, the location in the structure where the tested material or product is incorporated. In many instances, products are evaluated differently from materials. Products are often purchased with performance warranties and certification instead of specific qualification or test requirements.

17.1.2 Workmanship

The practices for determining compliance with the "minimum acceptable standards" definition are more varied than for specified properties. Where structural considerations are involved, such as with bearing value of piles, soil compaction, or the tightness of bolts, minimum standards of acceptance are well defined and widely published. Where more subjective judgments are involved, such as with the standard of workmanship for concrete wall finish, then common sense and experience should suffice. However, standards of measurement have been developed for most physical products of construction. If there are controlled minimum parameters, criteria to be met are specified in contract documents.

If minimum levels of acceptance for specified properties and workmanship are not identified, then the RPR relies upon typical industry standards. Table 17-1 lists some widely used elements of construction. Next to each element are respected national organizations with published information on standards of acceptance.

17.1.3 Statistical Quality Assurance and Quality Control

There is a trend in the construction industry toward statistical analysis for quality control by constructors and suppliers, and for statistically based acceptance criteria by owning agencies. The American Concrete Institute (ACI) publishes information on statistically based acceptance procedures for concrete. The Transportation Research Board has published several reports on statistically based specifications.

17.1.4 Requests for Substitution

Requests for substitutions of materials are common in the construction phase. Substitutions may be proposed by either party to the contract to save time or money or improve quality. It is necessary to obtain approval from the design professional for change requests. In all situations,

TABLE 17-1: SOURCES OF INFORMATION ON ACCEPTANCE STANDARDS

ELEMENT	SOURCE OF INFORMATION
Earthwork	American Society of Civil Engineers, Geotechnical Division ASFE—The Association of Engineering Firms Practicing in the Geosciences
Concrete	The American Concrete Institute Portland Cement Association Prestressed/Precast Concrete Institute American Society for Testing and Materials American Association of State Highway and Transportation Officials
Masonry	American Society for Testing and Materials American National Standards Institute National Institute of Standards and Technology Masonry Society Brick Institute of America National Concrete Masonry Association
Timber	American Institute of Timber Construction National Forest Products Association
Structural steel	American Institute of Steel Construction American Iron and Steel Institute
Reinforcing steel	Concrete Reinforcing Steel Institute Wire Reinforcement Institute
Asphalt	American Association of State Highway and Transportation Officials Asphalt Institute
Painting (of metals)	Steel Structures Painting Council
Electrical installations	Institute of Electrical and Electronic Engineers National Electrical Manufacturers Association National Electrical Code National Fire Protection Agency
Mechanical installations	American Society of Heating, Refrigeration and Air Conditioning Engineers American Society of Mechanical Engineers American Petroleum Institute American Water Works Association
Welding	American Welding Society Lincoln Welding Foundation American Society of Mechanical Engineers
Skid resistance	American Society for Testing and Materials

the requests for substitutions should be submitted and responded to in a timely manner.

Substitution of specified items requires a formal change order signed by both parties to the contract. On public-works projects, it is usually the constructor's right to offer substitutes, and, if they meet the specified criteria, the owner may allow them. Project specifications set forth procedures for proposing substitutes.

As with all factors bearing on quality, it is essential that the job record accurately reflect the item substituted, the original item, the reason for substitution, date of action, and whether a price adjustment was negotiated as a result of the change. The RPR maintains communication with the design professional and constructor to document the substitution and adjust the contract price, if necessary.

17.2 COST ESTIMATES AND PAYMENTS

Reliable estimates of cash-flow requirements necessary to maintain construction progress are vital to project owners. Providing timely and correct payment for work accomplished is critical for constructors. One significant task of the RPR is predicting project cash flow and arranging for appropriate payments.

Costs are divided into two broad categories: payments to the constructor and payments to others. Payments to constructors are further divided into regular or periodic payments of originally contemplated work, and extra work for change orders. In addition to payments to the design professional, payments to others fall into several categories, such as real estate and right-of-way acquisition, utility companies, testing laboratories, specialty consultants, vendors of equipment, and other construction-related costs.

A complete record of change orders, indicating the percentage of both change-order costs to each contract and the total change-order costs to the total project, is maintained by the RPR. Right-of-way, utility, surveying, and other inspection-agency fees should be coordinated with individual contracts, where appropriate, or to the entire project.

Periodic payments usually mean monthly payments to the constructor based on the value of work accomplished. On large projects, where considerable amounts may be expended in a given month, and particularly during periods of high interest rates, it is not uncommon to contract for more frequent payments based on the pace of work. For example, the contract may specify that payments are made at semimonthly intervals. On fast-paced projects with high labor costs, payments are sometimes made weekly.

There are occasions when a payment is made only once, when the project is completed. Furthermore, payments can be made when progress reaches a predetermined level, such as 10, 50 or 90% of completion. The timing of payments depends entirely on the project and the agreement between the two parties. However, for the majority of contract construction in the United States, monthly progress payment is normal.

The constructor is usually responsible for preparing the periodic-pay certificate, with review and approval of the RPR. On some public-works projects, however, the owning agency prepares the estimated value of the monthly pay certificate and the constructor reviews and acknowledges the amount due. The contract may establish the responsibility for initiating payment with either party. In the absence of any specific language in the contract, the party to whom the money is owed is responsible for initiating the request.

17.3 METHODS OF PAYMENTS

The method used for preparing the periodic payment certificate depends entirely on the type of contract involved. There are three principal types of contracts: unit-price, lump-sum, and cost-plus.

17.3.1 Unit-Price

Unit-price contracts are common in public works projects where the quantities of various kinds of materials and work segments are approximated and not precisely known. Thus contracts may be awarded calling for so many acres of clearing and grubbing, so many cubic yards of excavation, so many cubic yards of backfill, so many tons of pavement, etc., with a unit price bid or negotiated for each item.

Under such a contract, pay items are set up so that it is possible to measure the quantities involved for each item completed.

In all types of contracts it is sometimes necessary to pay for materials that were delivered but not yet incorporated in the work. For example, the contract may include the cost of reinforcing steel in the price per cubic yard of concrete. Delivering reinforcing steel is expensive and it is valuable property to the owner. In such situations, the value of the steel should be determined for payment before it is cast in place in concrete. The value of the material usually is determined by requesting copies of invoices of the materials.

Occasionally a situation arises that necessitates changing a unit price. Unforeseen circumstances, such as unknown soil conditions, or changes which greatly increase or decrease the amount of materials to be used, may require a renegotiation. The new price may be more or less than the original price, but it should reflect such factors as restocking charges, overhead amortization, and supplier's discounts. A contract duration change also may be necessary.

Partial payments are often made on uncompleted unit price items. The important point to remember is that all work has value, and estimating and recommending payment for that value under terms of the contract is a fundamental responsibility of the RPR. Partial payments are also made for materials which have been delivered to the job site but not yet incorporated into the work. Measures of value of partially completed work include not only the effort or funds expended to date but also the cost to complete. The effective transfer of control of the materials and matters of security also influence the value of partially completed work.

The RPR is responsible for recommending payment for only the stated value of a completed item, less the cost to complete it.

Accurate record-keeping, and an awareness of the value of materials are essential in dealing with unit-price contracts. This process is facilitated by periodic assessment of the value of completed work through joint review by the constructor and the RPR.

17.3.2 Lump-Sum

In lump-sum contracts, quantities of material and work-hours are determined by the constructor. The constructor submits a single lump-sum price for the completed facility. Typically, the successful constructor divides the contract into various components (schedule of values), similar to a unit-price contract. Mobilization or contract initiation costs are paid as a separate item, if permitted under the contract.

The RPR has the responsibility to determine if the various items of work included in the lump-sum breakdown are properly balanced to avoid overpayment for completion of early items. This is referred to as "front-loading." For example, foundation excavation precedes roofing on a contract. Some constructors would overstate the costs required to complete the foundation excavation and understate the costs of the roofing, since this would cause greater cash flow at the beginning of the project. Such a scenario would unfairly benefit the constructor and unfairly penalize the owner, even though the total final amount of the contract was not changed.

17.3.3 Cost-Plus

In cost-plus contracts, the constructor is reimbursed for actual costs plus an agreed-upon rate for overhead and profit. Because the constructor is compensated for costs rather than for completed work, the emphasis on record-keeping shifts from the amount of work completed to the costs for the completed work. Under this type of contract, record-keeping is more important and it is necessary to record each worker (direct, indirect, and supervisory), the hours worked, the type of work, and the wages paid. Equipment usage and costs are also recorded. Moreover, methods must be established to record and file the large quantity of material invoices, delivery slips, and other records required to verify the cost to the constructor to complete the work.

Complexities occur when cost-plus work is performed within a unit-price or lump-sum contract, such as when extra or unexpected work is encountered for which no unit price has been established. In such situations, the mixed use of personnel and equipment can be reimbursed to the appropriate pay item on a cost-plus basis.

17.4 RETAINAGE

Retainage is withholding earned funds, usually a percentage of the work completed to date, in case an error in quantity estimating, a lapse in meeting quality standards, or a construction error is discovered. It is important to understand that retainage is neither a penalty nor a license to alter the contract. In some cases, retainage is used as an inducement to encourage timely completion. However it is used, retainage is a temporary assessment against earned funds, to be released promptly after the cause of the assessment has been satisfactorily addressed.

Releasing earned funds fully and promptly while withholding all unearned funds is a challenge to competent contract administration. Some projects allow the posting of securities or other items of value in lieu of retainage, to be placed in escrow under the owner's control, or provide other methods to allow the constructor to earn interest on retained funds. Such escrow account will not be released until the project is accepted and final payment made.

17.5 LIQUIDATED DAMAGES

Liquidated damages are intended to compensate the owner for costs incurred and loss of income because the project is not substantially complete within the time specified in the contract documents. Liquidated damages can be assessed only when the cause for the delay on a project can be attributed to acts or omissions by the constructor. If a project is delayed for reasons beyond the control of the constructor, then sufficient extensions of time are granted under a contract-change order.

17.6 BONUS CLAUSES

Bonus clauses (incentive clauses) are categories of cost incentives. Bonuses may be prescribed for progress determinants as well as for quality determinants. When used as progress incentives, they should not be confused with liquidated damages. A bonus is not necessarily related to actual benefits as a result of finishing the project early or late. They are a predetermined sum proposed solely as an inducement, based on enhanced value received and defined by contract.

Bonuses for quality determinants usually are based on some statistical evaluation of a measurable quality attribute (the smoothness of pavement, the strength of concrete, or the density of compaction). Such clauses are an excellent means of rewarding constructors for high quality work.

17.7 CHANGE ORDERS

Construction usually involves creating custom-made products in the field. Consequently, the variety of foundation types, the weather, the nature of materials, design, fabrication, and erection frequently dictate some deviation from the original plan as defined by the contract documents. Prompt recognition of the need to change saves both owner and constructor from unnecessary cost increases and schedule delays.

All construction projects need contingency plans for unforeseen circumstances. Changes should be recognized in sufficient time for materials, designs, and fabrications and installations to be altered, estimated, performed, received, and fair prices negotiated. Often this is not possible, and the work must proceed before prices have been agreed upon. The RPR should proceed expeditiously to put a change order in place. Reaching a reasonable agreement is a prime responsibility of the RPR and the constructor. The goal is to reach an objective conclusion about a set of physical events with or without quantifiable costs. The inevitability of change, and the likelihood of disagreements as to the cost of the change, should be recognized and procedures developed for a satisfactory resolution. The contract documents may specify a procedure for calculating the dollar amount for an agreed-upon change.

In addition to changes to the original materials, design, or fabrication, it frequently becomes necessary or desirable to perform extra work on a project. This may entail providing more or less of an item than originally intended, and utilizing the skills and resources of the constructor to perform work or implement a concept not originally planned. In such situations, a change order is necessary.

The change-order documents accurately reflect the nature and reason for the change. The documents should be signed by the owner and the constructor after review by the design professional. Documents should be numbered, dated, and include relevant information, such as revised-plan sheets, sketches, specifications, and quotations. The document should address impacts of the changed work on the project schedule, when appropriate. It should be recognized that some changes are not within the scope of the original contract and the constructor should have the right to reject such changes.

In addition to changes to the original work scope and extra work, change orders are used to acknowledge changes in progress factors even where no physical change is evident. For example, if a contract schedule or a completion date is to be extended because of unusual weather, then a change order should be prepared reflecting that situation.

On most projects, particularly renovations or other projects where there is a potential for unforeseen conditions, the owner is well advised to budget a reasonable amount for change orders as a contingency cost. This type of provision helps the owner face additional costs if changes due to unknown or unforeseen conditions do occur, and may reduce disagreement and possible litigation among team members.

17.8 NONCONSTRUCTOR INVOICES

The certification and recommendation for payment of various vendors' invoices is an important element in effective project cost management. All such invoices—whether for utility relocations, purchase of equipment by the construction organization, or the testing of construction components—must show the date that the purchase was made or work was performed, the unit prices or costs involved, and the specifications or other quality criteria used in performing the work. Each vendor's invoice should be considered a separate contract and should provide the same information as in the construction contract.

17.9 CONSTRUCTION PROGRESS

Contract documents generally require the precommencement submittal of a progress schedule by the constructor. The schedule is important to the RPR in establishing cash-flow requirements, assessing personnel demands, and coordinating contract work with adjacent activities.

There are numerous ways to communicate the intent of progress. The most common kinds of schedules are bar charts (or Gantt charts), network analysis such as critical-path method (CPM), and "S" curves which relate progress to cumulative costs.

Bar charts are the simplest schedules to prepare and evaluate, because they show time on the horizontal axis and the various items of work on the vertical axis. Bar charts can become more detailed and complex simply by refining the time scale. The use of weeks rather than months, or days rather than weeks, greatly improves the detail.

A schedule could show bridge construction as a single element. Or it could show bridge construction as concrete in bridges, steel in bridges, pile driving for bridges, bridge surfacing, etc. A schedule could show each item of work for each pier and abutment in a bridge, such as pile driving for pier 1, concrete in footings for pier 1, reinforcement steel for footings in pier 1, etc. Thus, the level of detail required determines the schedule's complexity. The degree of detail should be outlined in the contract documents along with the criteria to be used, if any, for approval of such schedules.

In addition to the passage of time, network analysis recognizes the interrelationship of various construction elements. The elements of work are represented by an arrow symbol, which is generally undimensioned. The nodes, representing the tail and head of the arrow, can be tabulated for early or late start, or early or late finish, to present a realistic picture of the variability that is common to construction. As with bar charts, however, network analysis is not inherently simple or complex. The complexity derives from the number of elements chosen to be reflected. The information developed through network analysis can be effectively displayed in bar chart form.

"S" curves are prepared by combining the values of the elements of a schedule for the period of time shown on the schedule. These periodic totals are accumulated to show the typical rising "S" pattern curve to predict job progress. Although such curves usually are prepared by accumulating dollars, they also can be created by accumulating working hours or any other element of construction.

On larger projects, the owner may wish to consider a periodic independent review of the progress schedule so that an impartial assessment of any delays can be made while the project is ongoing.

17.10 PROGRESS REPORTS

Progress reports are issued to interested parties to communicate the status of construction as compared with a standard or norm, usually the progress schedule discussed in the preceding section. Although there are various reporting formats, the three common types of construction reports are detailed reports, summary reports, and subjective reports.

Detailed reports are prepared on a scheduled basis (usually daily), and involve tabulation of each item of work accomplished during the defined period. Detailed reports form the substance of the contract administration file and are an important resource for paying requisitions, resolving disputes, and recreating the job history. The preparation and review of detailed progress reports is a significant task for the RPR.

Summary reports contain information from each of the detailed reports and relate that information to project requirements. Summary reports may be made for any time period, but usually are done monthly. They convey information to all members of the project team and illustrate the amount and type of work accomplished during a given period of time. Prompt and knowledgeable review of such data is vital to maintaining project progress.

Subjective reports (exception reports) are not made periodically, but are filed when unusual or significant events occur during construction. For example, the summary report described in the preceding paragraph omits a detailed explanation as to why certain work was not performed during a specific period. The subjective report communicates why such deviation from schedule was necessary. These reports are filed in letter form and allow the RPR to advise all interested parties of significant events.

17.11 COMMUNICATION—CORRESPONDENCE AND RECORDS

All of the preceding sections on quality determinants, cost reports, and progress reports involve communication, a critical skill for the RPR. In this chapter, communications can be divided into two broad categories—written communication and records.

17.11.1 Written Communication

Written communication may comment, advise, recommend, or observe on various aspects of the construction operation. Communication may be recorded in the form of an exchange of letters between participants, reports—including progress reports, minutes of meetings, memoranda to the file, and written summaries of telephone calls and other oral communication.

The RPR is responsible for maintaining subject matter and chronological files of relevant correspondence and other written material.

17.11.2 Records

Other important records not previously discussed are part of the job record. These are the design and shop-drawing logs, job photographs (which are filed with the progress reports), and certified payroll records on projects where they are required by a federal or other government agency. In some jobs, nonconformance reports, which detail differences or discrepancies between "as designed" and "record," are prepared and therefore must be filed. Other documents, such as contract-change orders, which often are used to resolve nonconformance reports, are maintained and filed.

The shop-drawing log shows a number identifying the individual drawing, drawing title, the date it was received, to whom it was forwarded for review, the date it was returned, and the approval status. Job photographs should have some identifying date and number photographically developed as part of the print, showing when the photograph was taken. That number is referred to in a file explaining who took the photograph, the direction the photographer faced, and the activity reported to be shown by the photograph. These precautions are necessary for a potential legal challenge to accuracy.

Bidding documents are included as part of the permanent contract record. These include plans, specifications, bonds, and other similar affidavits that may have been required when the contract was awarded. Finally, record documents (sometimes inappropriately referred to as "as-built documents") showing revisions and additions to the original plans and specifications are maintained as a part of a proper job record.

17.12 CERTIFICATES OF COMPLETION

When a project is completed, many agencies require some sort of release or affidavit, or both, certifying that work has been done substantially in accordance with the contract documents, and that no outstanding payments are due. Furthermore, information on the location or completeness of record drawings may be required. Some agencies require lien or bond releases or additional guarantees as to nonpayment of illegal fees, bribes, or kickbacks.

The RPR is responsible for collecting and presenting these various releases and affidavits. The RPR's work continues well beyond the physical completion of construction.

CONCLUSION

The RPR is responsible for the administration of the owner's contracts involving activities during the construction phase of the project. In addition to performing the owner's duties and responsibilities assigned under contracts, the RPR is also responsible for building the project record. In this latter activity the RPR is assisted by the constructor and the design professional in the collection of data, preparation of reports, review of construction contract submittals, maintenance of "record" documents, and other activities defined in the owner/design professional agreement and the construction contract documents.

In discharging their responsibilities, RPRs place emphasis on:

- Maintaining quality of materials and workmanship.
- Considering and taking timely action on the constructor's requests for sub-

stitutions under the construction contract.

- Maintaining current estimates and making timely payments under the terms of contracts and agreements.
- Monitoring construction progress and maintaining current schedules and work activity charts.
- Revising change-order proposals and making timely decisions to approve, modify, or reject.
- Building the project record, including all forms of written communication.

In addition to contract administration as it applies to the owner's contracts and agreements, each project participant has responsibilities to assist in building the project file, records, and data base. These activities include preparation of reports, contract submittals, schedule preparation and update, records of material testing and equipment performance, and other activities to meet contract terms.

CHAPTER 18

START-UP, OPERATION AND MAINTENANCE

INTRODUCTION

The operational characteristics and maintenance requirements of the project after completion and turnover to the owner determine success in meeting project requirements. Operation and maintenance (O&M) factors influence life-cycle costs, continuity of service, durability, public health and safety, environmental impact, and other features of the completed facility. Consideration of O&M requirements in each phase of project planning, design, construction, and start-up is desirable.

The owner is responsible for O&M after the facility is completed. To implement quality operation and maintenance activities, the involvement of experienced operators is desirable. To this end, the owner may designate an O&M representative to advise and assist the design professional and constructor in planning, designing, and constructing the facility with O&M consideration in mind. In the planning and design of the project, the O&M concern is with input to and review of design-phase activities; in the construction phase, the concern is with construction observation and inspection; in the start-up phase, the concern is with verification, testing, and acceptance; and in the operational phase, the concern is with the operation and maintenance of the constructed project.

This chapter outlines procedures for large, complex projects where an experienced operating staff and experts in maintenance are available to advise at all stages of the project. Not all projects require this level of effort. For less complex projects that do not have such a staff, the same type of advice is given by experienced design professionals, equipment manufacturers, suppliers, plant operators, utility superintendents, and others having practical knowledge in the operation and maintenance of equipment and projects.

18.1 PLANNING FOR O&M INPUT AND TRAINING

The owner, when making contractual arrangements for the project, may select from a number of options in providing for consideration of O&M problems as they influence design and construction.

- The owner may appoint a member of the O&M staff as project coordinator to advise the design professional and the constructor from the O&M standpoint. For this assignment the owner should find an experienced individual, preferably a candidate to head up the operating staff for the completed facility.
- The owner may contract with the design professional to have an experienced member of the design professional's staff or a qualified consultant provide the appropriate O&M advice and review.
- The owner may delegate members of the O&M staff to work under the RPR or the design professional in observation and/or inspection of construction activity during the construction phase. This assignment provides an opportunity for O&M personnel to become familiar with the project while performing construction phase duties.
- The owner may delegate members of the O&M staff as well as the O&M coordinator to work with the design professional and constructor during the start-up phase of the project.
- The owner may contract with the design professional and/or constructor to provide review of and advice for operation and maintenance programs for some defined time after the project has been taken over by the operating staff.

Specific roles of the design professional, constructor, and O&M coordinators or advisers are defined in the owner/design professional agreement and the owner/constructor contract as those roles are influenced by O&M considerations.

18.2 DESIGN PHASE

In preliminary design, decisions are made relating to site selection and access, process choice, equipment selection, and other elements which impact operation and maintenance of the completed project. Since decisions made here limit flexibility in subsequent phases of the project, the O&M coordinator and adviser are consulted for choices on brands or models of equipment to be selected, arrangements of facilities, access for equipment repair, and other design features which influence O&M costs and activities.

Reviews stressing the operability and maintainability of various features of the project are scheduled at appropriate points in the design phase and at final design. The frequency and depth of these reviews varies with the size and complexity of the proposed facility. In the simplest case the owner and the design professional may informally review O&M features of the design. For more complex projects, special review teams consisting of the O&M coordinator and specialists in operations, together with the design professional and design specialists, may be formed to review and evaluate the O&M features of the proposed design.

Reviews from an O&M perspective normally include:

- Physical plant considerations, including size and layout of working space to be provided; suitability of equipment types, including efficiency in operation, maintenance schedule, and costs for the equipment; provisions for bypassing and isolating equipment for maintenance; specialized services, such as laboratory and chemicals; staff amenities, such as cafeteria, meeting rooms, and shower facilities; adequate lighting and ventilation; future expansion requirements; modular expansion possibilities; efficient land utilization; specific layouts of equipment, process and control systems to provide O&M accessibility; proper location of hoists; access for moving materials; laydown space and removal paths; appropriate flexibility and redundancy in equipment and controls; and provisions for adequate manufacturer-supplied materials, training, and spare parts information.
- Control strategies, including manual backup controls proposed for use and the effects of alternative strategies on efficiency of operations and staffing.
- Life-cycle cost considerations, including building materials and equipment.
- Environmental considerations, such as provisions to mitigate odors, noise and undesirable esthetic effects, as well as the possible need for a public-relations program.
- Safety considerations, including equipment, chemicals, protective devices, sprinklers, clothing, staff training.
- Personnel and budget planning for O&M staffing of the proposed facility, a tentative budget and staffing plan.

During the design phase, the owner is responsible for communicating needs, constraints, expectations, and requirements regarding performance, operation, and maintenance of the proposed facility, and for providing timely reviews. The owner is also responsible for providing adequate O&M input and determining (with the help of the design professional and the O&M coordinator) O&M budget and staffing requirements, and for initiating the hiring of supervisory personnel during the latter part of the design stage.

The design professional is responsible for preparing plans and specifications incorporating O&M considerations. The design professional includes in the construction contract documents provisions for equipment performance criteria, repair and replacement warranties, and adequate manufacturer-supplied O&M manuals, spare-parts information, operator training on new or complex equipment, and equipment start-up. The design professional is frequently authorized to prepare an O&M manual to include items such as process description, design criteria and equipment data, equipment purpose, operating parameters, potential problems and solutions, emergency operating procedures, safety, and other information. Plans and specifications, including backup documentation prepared by the design professional, constructor, or vendor are made available for O&M use.

18.3 CONSTRUCTION PHASE

As the design phase ends and construction starts, the owner and O&M coordinator are responsible for developing the O&M staffing plan and budget and for assembling the owner's O&M staff. The goal is to have a complete and trained staff before start-up and owner acceptance of the facility. The design professional provides assistance to the owner in the planning, budgeting, and training for O&M activities. The design professional's more direct contributions in support of O&M activities include review of manufacturers' data for incorporation in the O&M manual, structuring and coordinating training programs when requested by the owner to do so, and preparing for project start-up in cooperation with the constructor and O&M coordinator.

The construction phase of the project provides an opportunity for the owner's O&M coordinator and staff to make the transition from the advisory and review activities of the design phase to more active roles during the construction phase. Activities contributing to project construction which can also provide valuable information and training for O&M personnel include:

- Inspection and witnessing the testing of materials and equipment at the manufacturer's plant.
- Observation of installation and testing of equipment by the constructor.
- Observation of construction site activity to gain insight on utility locations, electrical conduit routings, installation problems affecting O&M, and arrangements of project elements as they affect operational safety and ease of maintenance.
- Assistance to the constructor in assembling documentation required under the construction contract documents.
- Assistance to the design professional in preparation and review of the O&M manual.

These activities and others of a similar nature can be performed strictly as a training exercise, or they can be performed as members of the RPR's field staff, or as members of the design professional's field staff during the construction phase. If members of the O&M staff are to function as part of the design professional's staff, the owner/design professional agreement must be written to define clearly lines of authority and responsibility for this situation.

The constructor's responsibilities relating to O&M include:

- Assembling and forwarding to the owner information required by the contract documents on various pieces or assemblies of equipment, including manufacturers' warranties, operating instructions, and maintenance requirements.
- Maintaining a current set of revised plans and specifications, including the effect of change orders and other pertinent information to guide the design professional in preparing record documents for the facility.
- Coordinating with the O&M staff on delivery and storage of spare parts, tools, and equipment to be used for project O&M.
- Working with the O&M coordinator and design professional in planning for and conducting project start-up.

During the construction phase of the project, the constructor is responsible for the acceptance-testing of various elements as specified in the contract documents and for the maintenance of these elements until turnover to the owner for operation. The constructor's role in project start-up and turnover varies from project to project and is defined by terms of the owner/constructor contract.

18.4 START-UP PHASE

The purpose of start-up phase activities is to demonstrate that project elements constructed or installed by the constructor are in working order, and that the facility performs as planned by the owner and design professional. This activity gives the operation and maintenance staff the opportunity to become familiar with the project under the guidance of the constructor and design professional.

Project start-up and turnover can be as simple as cutting the ribbon at the dedication ceremony for a new highway. Start-up of an industrial plant or petro-chemical facility may require the organization and training of a start-up group composed of representatives from the owner, design professional, and constructor. The owner's O&M staff are key players in the start-up of any project.

18.4.1 Planning the Start-up Program

Responsibility for organizing and leading the start-up program is generally addressed in the owner/design professional agreement and the owner/constructor contract and may be assigned to:

- The design professional on projects where the design professional has responsibility for drafting the O&M manual and for training O&M personnel.
- The constructor on design-build or turnkey projects.
- The owner, on industrial projects where the owner may have furnished or specified the process equipment; on

projects using multiple constructor and design professional assignments for elements of the total project; on expansion or remodeling projects when joint occupation of the site requires close coordination of construction and O&M activities.

With responsibility for start-up established, the start-up team is assembled with representation from the design professional, constructor, and owner, with particular emphasis on representation from the owner's O&M staff. Activities of the team in planning for start-up include:

- Preparing and reviewing start-up programs and procedures.
- Determining construction completion status.
- Scheduling start-up activities.
- Planning for supervision of system testing and correction of deficiencies.
- Reviewing final inspection reports and project closeout submittals.

The interaction and exchange of information among the principal parties involved in the project may be outlined in a start-up manual, along with planning, scheduling, testing, and other activities planned by the start-up team. The start-up manual should be structured to fit the project. Simple, direct, and brief language is preferred. Aids such as forms, checklists, and tabulations are useful.

18.4.2 Start-up Activities

Project start-up activities demonstrate the integration of various constructed systems into a unified facility. Major systems that eventually form a building or other facility are grouped into the following categories:

- Structure, consisting of foundations, slabs, bearing walls, and frames.
- Envelope, consisting of roofs, curtain walls, and ceilings.
- Life safety and habitat support systems. This is further divided into two categories: mechanical systems provide water supply, waste disposal, heat transfer, conveyances, and fire safety; electrical systems supply power and are tied into the mechanical system.
- Process systems utilizing specialized equipment supported by the mechanical and electrical systems for manufacturing, refining, or treating products.
- Interior, or architectural details, consisting of habitable components, such as partitions, hung ceilings, floors, wall coverings, etc.
- Exterior, consisting of parking lots, pedestrian access, landscaping, storm water drainage, utility and transport systems.

Accomplishing a smooth project start-up requires the same effective planning and scheduling techniques as did earlier stages of the design and construction process.

Start-up activities are generally based on the premise that project elements completed by the constructor have met the material, workmanship, and performance specifications contained in the owner/constructor contract. The start-up activities are structured to:

- Determine that each component of the project is in working order.
- Determine that these components can be integrated to operate as a facility which performs as planned by the owner and design professional.
- Provide a means of training O&M personnel in the operation of each of the components and of the completed facility.
- Validate O&M instructions and manuals prepared by the design professional or others.
- Check the file of record documents (plans, specifications, manufacturers' operating instructions, maintenance instructions, etc.) for appropriate scope and detail.
- Serve as a vehicle for acceptance of the constructor's completed contract and turnover of the facility to the owner's O&M staff for operation, if so specified in the owner/constructor contract.

18.5 OPERATING PHASE

The operating phase of the project is the sole responsibility of the owner and the O&M staff. The participation of O&M staff members in advising, observing, and assisting during the planning, design, and construction phases of the project provides experience and training for operating and maintaining the project after acceptance by the owner.

During the first year of operation the O&M staff works with and through the constructor in seeking enforcement of manufacturers' equipment warranties and correction of any defects found in the constructor's work during the warranty period as defined under the owner/constructor contract. The O&M staff may also wish to consult with the design professional to request clarification and amplification of operating and maintenance manuals, to seek advice in fine-tuning project operations, and to ask for assistance in testing and evaluating performance for conformance to design criteria and project requirements.

CONCLUSION

Quality projects are operated and maintained to perform in accordance with design criteria and project requirements. In order to achieve these goals, the active participation of the owner's O&M staff and/or others experienced in project O&M is desirable during all phases of the project. Input by O&M staff or O&M consultants during various times may be summarized as follows:

- Planning Phase: Advise from an O&M viewpoint.
- Design Phase: Advise on formation of alternatives; assist in evaluations; suggest equipment layouts, spare part requirements, material requirements for O&M, etc.
- Construction Phase: Observe construction; inspect and test equipment; direct placement of spare parts, materials for operation and maintenance, etc.
- Start-up Phase: Participate as active members of the start-up team in conducting tests and trial runs, training O&M personnel, project turnover.
- Operating Phase: Accept full responsibility for operation and maintenance of the facility; consult with the constructor and design professional.

The complexity of the activities outlined varies with the project. On some projects, O&M input will come directly from owners or their appointed coordinators. On other projects, where the owner employs a large and sophisticated O&M staff, the owner's staff may offer specialized advisers for the project, write O&M manuals and procedures, and conduct start-up operations before accepting the project from the constructor. For all projects, active participation of the owner from an O&M viewpoint adds to the owner's understanding of the design criteria and the effort necessary to translate the design professional's contract documents into an operating facility meeting project requirements.

CHAPTER 19

QUALITY ASSURANCE/QUALITY CONTROL CONSIDERATIONS

INTRODUCTION

This entire document is concerned with the consideration of quality in the constructed project. Other chapters discuss broad aspects of responsibilities, team selection and organization, contract formulation, project management and administration, etc. This chapter is concerned with examining the project activities of the owner, design professional, and constructor specifically directed toward supplying quality services or providing quality construction.

For the purposes of this discussion, the following definitions are used:

- Quality Assurance (QA) comprises all those planned and systematic actions necessary to provide confidence that items are designed and constructed in accordance with applicable standards and as specified by contract.
- Quality Control (QC) comprises the examination of services provided and work done, together with management and documentation necessary to demonstrate that these services and work meet contractual and regulatory requirements. QA/QC programs previously formulated by the design professional and the constructor are modified to meet the specific requirements of the project as defined by the owner/design professional agreement and the owner/constructor contract and by applicable regulatory requirements.

19.1 GETTING PROGRAM STARTED—OWNER

Project-specific QA programs are the responsibility of the owner. While most owners are desirous of having a quality project, many will need assistance in stating project quality assurance requirements. Early involvement with the selected design professional often provides the assistance needed in stating these requirements, in defining services required, and in negotiating the agreement for project services. The QA/QC program as it affects the performance of the design professional is agreed upon during this phase. The QA/QC program expected of the constructor is formulated by the owner, usually with the assistance of the design professional at a later date, during the preparation of the construction contract documents. The requirements of the QA/QC program are often developed in the form of a written quality assurance manual.

19.2 DESIGN PROFESSIONAL'S PROGRAM

The design professional is responsible for formulating and implementing the QA/QC program for the design phase to meet requirements under the agreement for professional services. The owner may wish to review and approve the program and may require documentation.

19.2.1 Procedures

The design professional may have a QA/QC program already in place to cover normal activities. This base program is usually adapted and expanded as necessary to meet any special requirements imposed by the owner and required by the project.

The quality management procedures employed by the design professional are used to improve thought processes, clarify communications among team members, and to translate the concepts and mental images of the project in the designer's mind to physical structures and systems to be built by the constructor. These means include:

- Staffing the design effort with experienced personnel; implementing the owner's statement of project requirements for design and seeking clarification or amplification as necessary; confirming field conditions as they influence design; establishing and maintaining open lines of communication among design team members and with the owner.

- Preparing, reviewing and coordinating designs, drawings, and specifications among design team members.
- Scheduling special reviews and progress reporting as appropriate for internal control and as required by the QA/QC program.
- Forming review activities to include advisers on construction, operation, and maintenance, and design specialists not included in the day-to-day design activities.
- Arranging for peer reviews as specified in the professional services agreement.

Confirmation to the owner that the design professional meets his or her QA/QC responsibilities is provided through submittal of the progress reports, reports of project reviews, and appropriate documentation at the end of certain phases of the project in compliance with the design professional's quality assurance program.

19.2.2 Project Phasing

The design professional and the owner will benefit, and quality of the design effort will be enhanced, if the design progresses in a planned and orderly manner through a series of phases:

- Pre-project: This phase consists of the initial actions of the owner and the design professional. Relationships between the two parties and project parameters are defined. Agreements for professional services are negotiated. A project-specific quality control program is agreed upon during this phase.
- Schematic Design Phase: A written project program based upon the executed agreement with the owner is developed to guide the design through the next three phases. This is a document establishing design parameters, constraints, space and regulatory requirements, owner interfacing and communications, quality control program and procedures. Schematic design studies are made to define project outlines, and probable project-cost estimates are developed. In-house project-concept reviews and owner reviews are conducted. A report is made documenting the results of this phase and presented to the owner for review and approval.
- Design Development Phase: When the owner approves the schematic design report, design development drawings and outline specifications are prepared. During this phase regulatory agency and utility company verification of the design is sought. The probable project costs are estimated and in-house and owner reviews conducted. A report on this phase is also presented to the owner for approval. At this point, the design is "frozen" in order that the next phase can be carried out in an orderly, efficient, and coordinated manner.
- Construction Documents Phase: With the design "frozen," controls are maintained on design changes while preparation of working drawings, technical specifications, and contract documents is completed. Further reviews are conducted, necessary changes made, probable construction cost estimates are prepared, and contract documents are filed with appropriate authorities. Elements of the constructor's QA/QC program are mandated in the contract documents as they define standards and materials of construction, performance and testing of equipment, and other matters. Reviews, in this phase particularly, should involve construction specialists to assist in assessing constructability. Contract documents are submitted to the owner, and usually to legal counsel, for review and approval.
- Bidding or Negotiating Phase: This phase is relatively short, but very critical with respect to proper documentation and quality control procedures. It begins with the preparation and publication of requests for bids or proposals and ends with confirmation of proper execution of the agreement between the owner and constructor, including provision of required submittals, bonds, and insurance by the constructor.
- Construction Phase: During this phase the design professional has varying degrees of participation according to the owner/design professional agreement. The constructor assumes a primary role in fulfilling the QA/QC obligations as-

signed under the construction contract documents. The owner or the design professional, if authorized by the owner, assumes the duties of contract administration, resident project representation, and inspection.

The design professional's role in the project QA/QC effort during the construction phase includes the technical review and approval as provided by contract of constructor submittals under the contract, and documentation of performance and qualification tests and other duties as specified by the QA/QC program.

19.2.3 Design Reviews or Audits

Design reviews or audits are cornerstones of the design professional's QA/ QC program. They do not replace the ongoing checking process required of the design team to identify and correct discrepancies in dimensions, incorrect notes and references to details on other plan sheets, conflicts between plans and specifications, and other problems of this nature. Nor are they peer reviews, which are covered elsewhere in this Guide. The design review is an internal quality control procedure usually carried out by members of the design team and a review board selected for their expertise. Design audits, if implemented, are performed by individuals other than members of the design team. Design reviews or audits have the purpose of establishing the necessary levels of quality of the design by identifying unsound concepts, analyzing constructability of the project, eliminating "reinvention of the wheel," and assisting in interdisciplinary coordination.

The review or audit team size will vary with the size and complexity of the project; however, it should consist of a core of experienced senior professionals supplemented by other design professionals representing pertinent disciplines. By keeping the same core group membership for all design reviews, continuity in the review effort is maintained. Audit team members should be individuals not directly associated with the design.

The review or audit team usually receives appropriate material from the design team leader concerning the work for study prior to convening. The review or audit team and design team meet for discussion of the project and defense of the design concepts by the design team. It is important to document that these reviews have been held, but recording of discussions should be limited.

During the design-development phase additional reviews should be scheduled according to the needs of the project. As a minimum, a final review sometime around the 90% completion of the design should be held.

Reviews by the owner as provided for in the QA/QC program and design-development phases are treated as an opportunity by both parties to conduct a two-way exchange of information. The owner has the opportunity and responsibility to review reports submitted in draft form and to request that discussions be clarified or supplemented before the design professional's documents are submitted in final form for approval.

19.2.4 Quality Control During Bidding or Negotiating Process

The design professional's responsibilities during the bidding or negotiating phase, which may impact the quality and integrity of the bidding process, include:

- Completion of plans and specifications requires that the design team schedule their effort to allow proper checking, coordination, and design review or audit and approval before issuing documents for bid. Issuing large and often hurried addenda during the bidding period increases the probability of errors and omissions and creates confusion, frustration, and other difficulties for the constructors preparing bids.
- The pre-bid conference, which usually includes a job site tour, is an important introduction to the project for potential bidders. The primary emphasis of the quality control program here is to provide frank and forthright information about the project and the bidding documents. If questions are asked that indicate ambiguity or lack of appropriate information in the bid documents, it is not appropriate to issue clarifications or supplementary information at the conference, but only to indicate that pertinent addenda will be issued. Addenda are issued with receipt acknowledged by document holders and with sufficient time allowed to permit incorporation in the bidder's submittal.
- Bid analysis and evaluation requires careful review of bids received to determine that each bidder has complied

with the bidding documents. Spread sheets containing bid amounts by each bidder for each bid item, and tabulation of other requirements for each bidder, are useful in identifying irregularities in bids that may reflect lack of understanding or errors on the part of the bidder.

19.3 CONSTRUCTOR'S PROGRAM

The constructor is responsible for formulating, implementing, and administering the owner's QA/QC program required to properly fulfill the owner/constructor contract requirements. Additional program elements are formulated for the constructor's own interest in controlling the quality of the product.

In some ways the constructor's QA/QC requirements are easier to formulate and understand, since most contract quality requirements have to do with physical properties which can be defined and measured as opposed to the requirements placed on the design professional for maintaining quality in the concept and communication of ideas. On the other hand, the QA/QC program has become much more complex because the constructor is responsible for the activities of subcontractors, material suppliers, manufacturers, fabricators, and vendors, as well as for his or her own activities. In addition to the owner and design professional, a fourth entity, the public, is involved by virtue of building codes, toxic and hazardous materials handling, storage and disposal regulations, permits, and other regulations that impact the construction process.

19.3.1 General Elements

Constructors, before becoming involved in a project, will have their own QA/QC programs in place. Some of the procedures used include:

- Assigning qualified and experienced management and supervisory personnel to the project.
- Recruiting and assigning a skilled work force.
- Drafting purchase orders and subcontractor agreements to conform to quality requirements of the owner/constructor contract.
- Scheduling activities required to meet regulatory agency requirements, including necessary documentation.
- Performing periodic in-house review or audits of compliance with QA/QC program requirements.

19.3.2 Contractual Requirements

The owner/constructor contract documents represent the level of quality to be incorporated into the project by the constructor. They define standards and materials of construction, requirements for the execution of the work, performance criteria and testing of equipment assemblies and systems, and the documentation necessary to demonstrate that these contract requirements have been met.

The constructor's organization and execution of the quality control function in the preparation and processing of submittals required under the contract and the QA/QC program is essential to the smooth work flow on the project. Interruption of scheduled work because of delays by subcontractors, constructor, design professional, or owner in the contract submittal, review, and approval process is to be avoided. The constructor, assisted by subcontractors, suppliers, and vendors, is responsible for submittal of complete and technically accurate documentation as required under the contract. Attention to these requirements on initial submittals is essential to maintaining the schedules and procedures agreed upon with the owner and design professional who are obligated to perform review and approval in a timely manner. Recycling incomplete submittals interferes with activities of all team members and may create unnecessary conflicts.

19.3.3 Project-Specific Requirements

Project-specific requirements of the constructor's QA/QC program include:

- Use of qualified subcontractors (certified and/or licensed if required by the contract documents).
- Inspection and control of purchased materials, equipment, and services.
- Identification and storage of materials, parts, and components pending incorporation into the project.
- Control of measuring and test equipment.
- Segregation and disposition of nonconforming materials, parts, or components.
- Maintenance of records specified by contract and required by the constructor's own QA/QC implementation pro-

gram to furnish evidence of activities affecting quality.

CONCLUSION

The responsibility for specifying and funding quality assurance and quality control activity on the project lies with the owner. The owner, assisted by the design professional and other advisers, specifies requirements to be incorporated in the project contract documents.

During the design phases of the project the design professional is responsible for the formulation, implementation, and administration of the QA/QC program. Documentation is presented to the owner for review and approval.

During the construction phase the constructor is responsible for implementation of the QA/QC program, with the owner and design professional (when authorized) acting in the review and approval mode. Documentation required by the QA/QC program is submitted to the owner for review and approval. The design professional usually reviews technical elements of the documentation and advises the RPR.

Quality in the constructed project is much more than merely quality assurance/control programs carried out by the project participants. It is a state of mind of all involved in the project that places quality foremost. It requires a major communication effort during the entire process to keep all parties informed of the vital elements of the work and the concerns of the owner, design professional, and constructor. It also requires mutual understanding of those concerns and a realization that few, if any, construction jobs are without problems. Finally, it requires determination by all parties to resolve these problems equitably as they occur.

CHAPTER 20

PROJECT QUALITY THROUGH USE OF COMPUTERS

INTRODUCTION

Effective use of computers and computer systems assists design professionals, constructors, and owners when planning, designing, and constructing a project. Computer systems are capable of reducing the time required to perform many construction and design-related functions and, on many projects, computers and computer methods assist in improving elements of quality, including quality of the design and total life-cycle costs.

Computers are used in three basic areas:

1. Design applications.
2. Project coordination and data management.
3. Management and accounting, systems development, administration, and scheduling.

Computers, employed, managed, and maintained to their maximum, can give the project team considerable data processing powers, including:

- Analytical tools that allow simulation of the completed project's operation or performance to aid design decisions.
- Synthesis tools that allow automated selection of design variables using a volume and sophistication of computation not practical for manual methods.
- Graphics that allow visualization of the potential consequences of design decisions, rapid verification of input and output, explanation of design issues to nontechnical participants, and filing and retrieval of data used in the construction process.
- Shared data bases that allow prompt, accurate, and economical exchange of information among participants in the design and construction process.
- Access to generic data bases providing accurate, economical, and timely access to information required for design decisions.

This chapter gives guidance on computer usage to help realize the potential to provide quick and accurate computations, expeditious communications within the project team, fingertip data storage and retrieval, accurate design drawings and specifications, and facilitation of construction management activities/procedures.

20.1 EVALUATING FUNCTIONS

Typical considerations in selecting a particular function for computerization include:

- Does the function require complex mathematical calculations?
- How complex are the data-processing procedures?
- Does the task require storage and retrieval of large amounts of data?
- Can the data be stored and retrieved in standard formats that facilitate transfer and interpretation by different parties?
- Does reliable, economical, well-documented, "user-friendly" software exist?
- Do the benefits of computerization outweigh the several types of costs compared with acceptable alternatives?

If a computer or computer program is determined to be an appropriate tool to perform a specific function, the following questions should be answered:

- How difficult is it to develop, acquire, or customize the program?
- What is the best way to handle the data-entry function?
- Are error-detection features a primary concern?
- How easily can the program be learned by a new employee?
- What are the risks associated with malfunction due to power failures, breakdowns, and human errors?

The answers to these questions result in the decision to use a specific program designed for the function, or a utility application such as an electronic spreadsheet, or in the selection of any of numerous other programs that can be purchased for various custom uses.

20.1.1 Project Computer Systems

The greatest benefits of computers are achieved when data and analytical methods are planned and executed on a project-wide basis rather than as a series of disjointed tasks. Written procedures provide guidance to team members for operations within established criteria, for linking of hardware and assignment of responsibilities. If the project is complex, a procedures manual may be required. The overall objective is to be able to communicate information to all members of the project team in an effective manner.

20.1.2 Design Quality

Design professionals are well advised to use only those data processing and scientific application tools for which the functions and limitations are understood. Quality suffers, life safety may be threatened, and property may be damaged if the design professional is unsure of assumptions that form a computer program or of limitations that are imposed. Design professionals use appropriate analytical and design methods. "Appropriate" does not necessarily mean the most advanced or rigorous. Frequently, simple, less complex methods are appropriate.

20.1.3 Hardware and Software Requirements

Owners, design professionals, and constructors endeavor to select computer hardware and software with sufficient capacity and accuracy to meet the needs of typical projects being done in the office or in the field. Considerations for the appropriate hardware system and software programs are:

- Level of technology being used in projects.
- Level of communications desired by owner, design professional, and constructor.
- Flexibility in design.
- Sophistication of word processing.
- Compatibility with other systems.

In applications where team members are not experienced, experts in computer applications may be employed to give guidance on selection and installation of computer systems. The computer system is selected to meet the needs of the projects in a cost-effective manner. The system is structured to employ a mechanism of change control to determine in each case that the version of the software used is the one which has been tested and currently authorized for use. The rapidity with which software is enhanced or changed, particularly in the microcomputer field, makes it easy to confuse software versions. Revised versions may incorporate errors not present in previous versions.

20.1.4 Judging Software Results

Just as design assumptions only approximate reality, software models are only an approximation of reality. Except in trivial situations, different software employed by the same design professional, or the same software employed by different design professionals, may produce different numerical results. The results are influenced by the methods and assumptions made by the design professional employing the model. While many different analyses could be "correct," because there is seldom a single correct answer, there usually will be one model that best fits a given situation. It is the duty of the design professional to select the appropriate model.

20.1.5 Generic Data; Data Retention and Retrieval

Project computer data are an extension of the paper records of a project and are treated with the same care as would be accorded other documents. This may require transferring the data to a nonvolatile archival medium such as magnetic tape. In the absence of contractual relationships to the contrary, project design data remain property of the design professional.

Over the years, checking and sign-off procedures have been developed for the paper design documents of record for a project. Similar procedures are possible for electronic records. In a manner similar to manual calculations, input data are checked for validity and consistency. That check is not performed by the developer of the information, but by an independent checker.

Data should be accessible and secure during the course of a project. If the project is developed with all parties sharing and using the project data, precautions may be necessary to maintain accurate data. Newly created information is identified by its "owner," the person who can rightfully revise the data. All others are merely granted permission to examine and use, but not to change, the data.

Multiple versions of what are supposedly the same data are dangerous. Making multiple private copies of data is not recommended. When possible, only one "official" copy of the data should exist except for backup files. Obsolete data should be replaced with more current information. Outdated data may have to be retained for purposes of historical documentation, but should be removed from active access.

Software and data should have adequate maintenance and backup. Loss of critical computer software or important data may have a significant effect on the cost or completion time of a project.

Computer data are stored at the conclusion of a project. Careful handling is necessary to prevent loss of information due to aging or contamination. Furthermore, the data should be stored in a way that will allow retrieval and manipulation by a later generation of computers. Outdated versions of software are retained if the software is significant for the project archives. In this regard computer data are treated no differently from paper records of a project.

20.2 SPECIFIC COMPUTER CONSIDERATIONS FOR DESIGN

Computer applications are useful in activities associated with design effort, including project programming, conceptual design, preliminary design, and final design, as well as reconciliation of as-designed and as-constructed data.

20.2.1 Project Programming

Esthetic, functional, environmental, safety, and economic criteria are formulated for the project. Some may be qualitative and others quantitative in expression. Effective computer use in project programming includes:

- Search of appropriate data bases to obtain pertinent and current regulations the project must meet (zoning, building codes, safety, environmental, etc.).
- Use of data bases and design study techniques to define benefit-cost functions for the project and to establish values for design criteria.
- Consultation with other participants in the project, using automatic or manual methods of information exchange, to identify and respond to conflicts in criteria and opportunities for synergy.

Responsible computer use entails testing the criteria established. Site visits are particularly helpful. At the site of the project, special environmental conditions, natural or human, can be identified and assessed. At the sites of similar facilities, the effectiveness of project programs can be assessed and special requirements for the current project identified.

20.2.2 Conceptual Design

Conceptual design identifies solution schemes consistent with the program and establishes the additional design criteria required for each scheme. Effective computer use in conceptual design includes:

- Use of graphics programs to sketch solution schemes and assess their fit to the program.
- Access to data sources and knowledge systems to define scheme-specific environmental actions.
- Access to data sources and knowledge systems to identify critical failure mechanisms and reliable techniques for predicting the system response.
- Use of graphics systems to test the fit of subschemes among the project team, and to assist nontechnical participants in evaluation of alternative schemes.

Responsible computer use entails testing the validity of the criteria required for each scheme to be considered. Quantitative consideration of each scheme's ability to satisfy project criteria is deferred to preliminary design.

20.2.3 Preliminary Design

A few critical design criteria, variables, and actions usually determine the validity of a particular solution to the project. In preliminary design, these controlling instances of design criteria, variables, and actions are identified by the design professional, and values of the design variables are found (if possible) to meet the design criteria. The scheme is extended to detailed design only if preliminary design results are promising. Effective computer use in preliminary design includes:

- Accessing data and knowledge systems that provide simple, transparent models for prediction of performance.
- Applying manual, simplified, computer or other analytical techniques to establish trial values of design variables that will nearly satisfy design criteria.

- Use of computer programs or knowledge systems to study the characteristics of the design and the agreement with other design criteria.
- Exchanging preliminary design data with other participants in the project to identify and resolve inconsistencies.

Responsible computer use entails regular independent review of preliminary design results for validity and consistency. In most areas of civil engineering, traditional, rational, approximate methods of analysis are highly effective in testing the results of computer methods.

20.2.4 Final Design

Final design involves determining the value of each design variable, satisfying design criteria, and preparing data bases, plans, and specifications to transmit the results of the design professional's efforts to other project participants. Requirements for effective and responsible computer use include the following steps:

- Select analytical modeling techniques for their ability to input efficiently and test input, and to emphasize critical response results.
- Use data sources or knowledge systems to guide formulation of the analytical model for each subsystem, to achieve necessary accuracy economically.
- Evaluate critically the results of analysis with the corresponding results of the independent analytical method used in preliminary design.
- Use appropriate techniques for the design, considering factors such as reliability and life-cycle cost.
- Use computerized representations of the designs of the various subsystems. Employ automatic interference checking, when available, and review by the project team to identify and remove inconsistencies. (Production of drawings at a common scale, and overlaying to identify interferences, are effective tests for consistency.)
- Use computerized master specifications to formulate job specifications that are consistent with recognized practice for the type and location of the project. Review the results manually for credibility. Emphasize the special, project-specific elements of the specification so that these elements are not lost in routine application of master specifications. Conform the master specification to reflect any changes dictated by updated codes and standard use.
- Use computerized quantity takeoff techniques to prepare bills of materials in formats consistent with recognized practice for the type and location of the project. Test the results with independent manual or automated calculations.
- Provide data files on analytical models, bills of materials, plans and specifications to facilitate regulatory reviews and approvals, and preparation of bids. Check these with manual review of key drawings or views of the data.
- Establish procedures for updating data files on design and construction documents as changes occur during design, review and approval, contracting and construction.

20.2.5 Reconciliation of As-Designed and As-Constructed Data

In construction projects, deviations may develop between the contract documents and the as-constructed project. Such deviations are a consequence of field conditions which are different from those envisioned during design, and construction problems whose resolution results in a contract change.

Reconciliation of as-designed and as-constructed data may involve the development and implementation of a procedure to determine compliance with design documents by the material supplier, fabricator, erector, constructor, etc., and the review and approval of any necessary changes.

Using computers in their data-base management mode may help eliminate, record, and track design or field-changes to their resolution and incorporation into the project documentation.

20.3 COMPUTERS IN CONSTRUCTION—ADMINISTRATIVE USE

The functions that computer applications perform in the management of construction can be categorized into the following areas: corporate accounting, project management and administration, and special applications. This section discusses how project team members can use certain computer systems for administrative functions. A description of desirable features and available software is included.

20.3.1 Corporate Accounting

- General ledger. As the ultimate repository of all financial activity of the business, certain computer systems can integrate diverse accounting transactions into a single file, or data base, from which reports can be accessed. This is used to manage and analyze accurately the profitability of the company and provide reports to satisfy government audit requirements.
- Accounts payable. This system monitors open invoices, or bills, the company owes its creditors. It can be used to print checks and help to analyze cash requirements by department, vendor, job, age, or payment priority.
- Accounts receivable. This system monitors open receivables and can generate monthly statements, invoices, and aging reports.
- Payroll. Appropriate computer systems can produce checks and calculate withholding amounts, union deductions, and overtime amounts based on union agreements. It also permits allocation of employee time to various projects and produces year-end W-2s and government-required electronic tapes.

20.3.2 Project Management (Cost Control, Scheduling, Material Control, Contracting) and Project Administration

- Change-order estimating. This allows for the immediate identification of the cost impact of events potentially affecting the cost to complete the project. In a construction management environment, it can be used to identify and track costs as scope or field conditions change.
- Requisitioning. This system can produce formatted worksheets for the payment-application process. It can be implemented as a subset of the job-costing system.
- Accounts payable. This performs tasks similar to those described for corporate accounts payable. However, in a general or subcontracting environment the data flow to the job-cost and the general-ledger systems.
- Job-cost reporting. This is a data base file management system that receives data from the cost-control systems for timely monitoring of all costs. It produces reports with an end-user, free formatting report-writing feature, identifies cost to complete, and compares actual versus estimated costs.
- Cash-flow reporting. This is utilized to forecast, monitor, predict, and compare original versus actual out-of-pocket case requirements or earned value, or both. It accurately reports job status.
- Action lists. These are unrelated tasks that can be monitored in tabular format and accommodated in a spreadsheet format. Integrating listed and logical field activity is possible with scheduling software that provides for the reception of data from spreadsheets.
- Network-scheduling logic. This is used for planning, monitoring, and reporting. Systems are widely available in personal computer applications. Systems that provide custom free-format reports and easy-to-read bar-chart graphics are readily available.
- Quantity takeoff. This quantifies the scope of work indicated on design drawings. Blue-line drawings are quantified with an interactive interface, light pen, or digitizing tablet. Computer-generated drawings developed with a CAD system may generate bills-of-materials.
- Cost estimates. This allows the preparation of estimates that are uniform and free from mathematical errors. It maintains a pricing catalog with "standard" default values or special unit costs input by the estimator, permitting rapid recalculation of extensions.
- Bid solicitation. This system helps constructors select, notify, and solicit bids from appropriate subcontractors.
- Bid evaluation. This provides a comparative spreadsheet of the bids submitted. It allows for an item-by-item review of submissions by bidders.

- Control of materials. This tracks flow of materials from purchase to delivery to job site to installation.
- Letters. Word-processing packages automate the processes of creating, storing, sorting, retrieving, and merging large quantities of information. These packages can maintain contracts, standard agreements, and office forms, and can be used for name and address management.
- Shop-drawing control. This monitors all submittals required on the project, which submittals have been made, and the status of each submittal.
- Forms and agreements. These word-processing systems and advanced laser printers provide a complete and convenient inventory of standard agreements, office forms, payment requests, etc.
- Desk management. This system automates the storage, retrieval, and processing of large volumes of information relating to clients, consultants, constructors, and other organizations that are involved on many constructed projects. It provides fast retrieval and mailing-list capabilities with sort-and-search functions.

Computer systems can be used effectively in all phases of the project by each of the team members to produce timely and accurate reports and to enhance the communication process.

20.3.3 Special Applications

In addition to administrative functions, computer systems can assist team members in many types of projects, such as:

- Telecommunications. Special computer programs can easily connect remote sites and permit access to time-sharing systems by use of modems or multiplexers and dial-up telephone lines.
- Computer-aided design (CAD). This permits planning, cost estimation, cost allocation, space allocation, and maintenance planning. A CAD system maintains a data base of standard planning elements that can be incorporated into design. Summary and detailed lists of equipment and material quantities can be made quickly.
- Income property analysis. This produces a complete financial package (pro-forma) that can determine the financial feasibility of a proposed project.

CONCLUSION

Effective use of computers and computer systems can assist owners, design professionals, and constructors when planning, designing, constructing, and operating the project. Computer systems help to reduce the time required to perform many construction and design-related functions and, properly managed, computers and computer methods can assist in achieving quality.

Computers are influencing design practice by greatly increasing speed and accuracy of computation and increasing options (alternatives) considered. Indeed, effective use of computer methods may become essential to remaining economically and technically competitive. Computer methods are used responsibly when design professionals maintain professional control of their decisions, understand the technical bases for those decisions, provide adequate training, and independently evaluate significant data upon which the design decisions are based.

Specific computer programs can perform many time-consuming and administrative functions quickly and efficiently. For example, computers allow construction personnel to monitor accurately and rapidly changes in contract commitments, costs, schedules, impact of change orders, subcontractor relationships, material deliveries, and many other aspects of complex projects. As the job progresses, profit and loss can be monitored. Resource productivity can be greatly improved, and potential problems can be identified early. Organizational efficiency can be fine-tuned, job estimating can be made more accurate, and out-of-sequence work curtailed. Computers can also improve communications and teamwork within the project team by providing access to project information, helping the project team to function with greater efficiency.

CHAPTER 21

PEER REVIEW

INTRODUCTION

What Is a Peer Review?

Peer reviews have been utilized successfully for a number of years to advance quality in organizations and on projects. Many believe that increased use of peer reviews would improve quality overall in the construction process, and that owners, design professionals, and constructors should carefully consider utilizing peer review for appropriate projects. This chapter discusses the two most significant categories of peer review: organizational peer reviews and project peer reviews (see Figure 21-1).

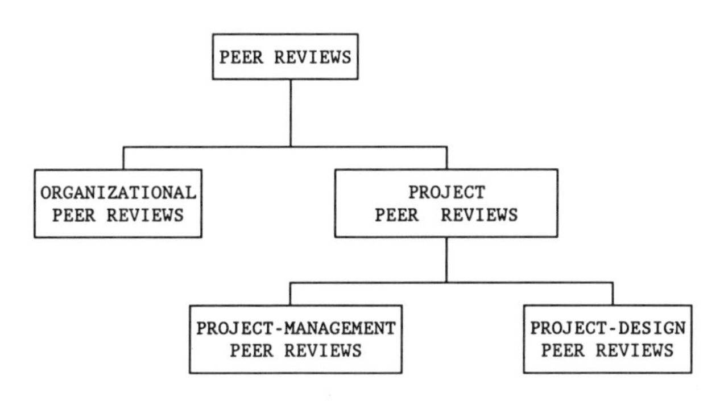

Fig. 21-1. Categories of Peer Reviews

Organizational peer reviews address organizations as a whole, focusing on policies, procedures, and practices, and not on any single project. Project peer reviews focus on particular projects, not on an organization's practice in general.

Project peer reviews can be further divided into two basic categories, i.e., project management and project performance. Project-management peer review concentrates on the same policies, procedures, and practices as would be examined in an organizational peer review, but with a narrower scope of applicability to enhance the quality of management on one particular project. Project-performance peer review examines in detail the technical results or recommendations proposed for, or the construction methods employed on the project at the time of the peer review, usually with limited regard for the management of the process that led to the technical documents or procedures being examined.

What Is a Peer Reviewer?

A peer reviewer is an experienced professional who has performed similar functions in similar situations; often a manager or senior technical person from another organization or separate division of the same organization. Typically, reviewers are recognized for their expertise and contributions to practice, and are active in their respective professional or trade associations. They are generally trained in the peer review process to sharpen their objectivity and enhance their communication and listening skills.

Purpose of Peer Review

The purpose of peer review is to enhance the quality of the constructed project by bringing additional independent, high-caliber expertise to examine the planning, design, and construction processes. The review may address the quality of the technical design as expressed in the contract documents, the management of a project, and/or the management of organizations performing those functions, depending on the scope desired by the owner and the construction team.

21.1 GENERAL BACKGROUND

In the context of design and construction, a peer review is:

- Recognition that an independent team of peers can bring a new range of experience and opinions to the organization or project.
- A review of an organization or project, conducted by peers of the original owner(s), manager(s), author(s), design professional(s), or constructor(s) who are independent of the subject of the

review. A peer is defined as a person or group of persons with the same or higher level of technical or managerial expertise as those who are responsible for the subject of the review. Reviewers are generally from a separate organization, to eliminate relationships that would interfere with the impartiality of the review. If the reviewers are from within the same organization that is responsible for the office or project being reviewed, as occurs with some large organizations that have formal internal peer review programs, the reviewers should be sufficiently remote geographically and administratively so that there is no question of their independence and objectivity.

- Seen by all participants as an endeavor that demands special attention and procedures because special benefits are expected to accrue.
- Addressed to a specified purpose, scope, format, and duration, distinguishing it from other reviews that may be performed on a less systematic basis. It has a commissioning, a beginning, a report, and a decommissioning.
- Intended to produce results. Therefore, the recommendations should be disseminated as necessary to bring about implementation, especially to the persons whose activities were the subject of the review. In the case of a project peer review, the owner, the design professional, and/or the constructor each retain authority and responsibility for their contractually agreed-upon project activities.
- Paid for by the commissioning authority, which benefits from this valuable service. If the commissioning authority is the owner, the review improves the project or gives the owner additional assurance that project requirements will be met. If the commissioning authority is the design professional or constructor, it improves the quality of performance.

A peer review is NOT:

- An indication that the owner, design professional, or constructor is incompetent or suspect in any way. It is simply recognition that an independent team of peers can bring a new range of experience and judgment to the organization or project.
- A review by a building code official or by any other governmental agency as it carries out its regularly mandated regulatory responsibilities covering some area of the design or construction processes.
- A constructability review, although a peer reviewer might ask the question, "Was a constructability review performed?"
- A value-engineering study, despite certain similarities in the two processes. Value engineering of a design focuses on potential cost savings, while a peer review of a design examines the quality of the design. A peer reviewer is likely to ask first, "Were the right procedures followed?" whereas the usual value engineering question is, "Is there money to be saved?" The questions are not mutually exclusive and can be merged into a broader inquiry.

Other features of peer review include:

- Confidentiality. Often reviews are confidential in-house events, while others dealing with public projects may lead directly to public reports. Reports may be either written or oral, and be either technically oriented or for general management understanding. In either case, confidentiality between persons interviewed and reviewers is extremely important to the integrity of the process.
- A peer review may be voluntarily requested and authorized, or it may be mandated by some authority other than the agency to be reviewed.
- Size of the organization or project is not necessarily a deciding factor in commissioning a review, although a large project might be a candidate for a peer review primarily on the basis of its size.
- To be successful, a peer review requires adequate resources in terms of budget, time, and effort. Otherwise,

the exercise could lead to incomplete, misleading, and unsatisfactory results.

21.2 BENEFITS OF PEER REVIEW

There are many benefits derived from peer review which relate to enhanced quality, both for the project and for the organizations forming the construction team.

The first benefit comes in the form of awareness: a sensitivity to the fact that a project is important enough to warrant a peer review, or that management cares enough to have its organization peer reviewed. The preparation for a review often uncovers neglected areas and focuses attention on them in advance of the review.

Peer review leads to a sharing of expertise which enhances the construction process. It offers the opportunity to identify problems and take corrective action in time to have a positive impact or effect.

21.3 TYPES OF PEER REVIEW

21.3.1 Organizational Peer Review

Organizational peer review assesses a design or construction organization's operations in light of its overall policies and practices, leading to recommendations for general improvements. Variations on that overall scope do exist. For some organizational peer reviews, the emphasis may be on whether the policies and procedures that have been established are satisfactory. In other cases, an organization may have well-established policies and procedures, and the purpose of the peer review is to determine if they are actually being practiced. The organization to be reviewed may be private, public, or a combination of both.

21.3.2 Project Peer Review

Project peer review is a separate, structured, comprehensive, and thorough fact-finding process conducted by one or more senior professionals who are separate and independent from the project team. The scope is defined in detail by the reviewer or review team with the design professional, constructor and owner, before the review is started, and includes the functions to be reviewed, the process to be followed, the schedule, and the form of reporting.

Project peer review goes beyond the normal checks performed within a design or construction organization that are part of standard procedures and everyday routines associated with quality management.

21.3.2.1 Project-Management Peer Review

Many industry professionals believe that more deficiencies result from mistakes in the management of a project than from purely technical errors. A project-management peer review is intended to address this particular problem. The specific scope of the program is to review elements that directly reflect the management of the process as contrasted to a technical review of design assumptions, calculations, construction documents, or construction methods, and adherence to regulations. A project-management peer review is somewhat similar to an organizational peer review in emphasizing the process followed, but narrows the examination to a single project.

The intent of a project-management peer review is to assist and improve the performance of project management. The review may be commissioned by the owner, designer, or constructor. In each case, the detailed scope of the review is subject to variation, and therefore should normally be defined in writing by the commissioning agency with agreement by the peer-review team.

Projects or segments of projects having one or more of the following characteristics are candidates for peer review:

- Extreme size and/or complexity.
- State-of-the-art expertise requirements.
- A project type that has experienced previous difficulty.
- A project whose nature has changed during design.
- A diverse team (e.g., several offices or organizations involved).
- A hurried design schedule, including the fast-track approach.
- Limited budgets in terms of cost and staff effort.
- Severe staffing difficulties.
- A high potential liability to the owner, design professional, or constructor, or a high risk to the general public.
- A change in project managers.
- Doubts as to the true status of work performed to date.
- An environmentally sensitive project.

21.3.2.2 Project-Performance Peer Review

A project-performance peer review is a comprehensive examination of the technical aspects of the project design as they relate to the concept, progress, or final results, or the construction means, methods, and techniques employed. The review may be performed at various stages during the course of the design or construction effort, or only upon completion of the project (although the latter will really only help the next project). The scope is defined in writing at the time a review is authorized and may typically include an inquiry into:

- Design assumptions.
- Applicable codes and regulations.
- Accuracy of calculations.
- Appropriateness of selections from alternatives.
- Application of judgment.
- Constructability of the project.
- Construction means, methods, and techniques employed.
- Feasibility of meeting the project requirements.

Guidelines for determining whether a project should be considered for project-performance peer review vary considerably geographically and among organizations within the same locality. Therefore, a list of considerations is presented to facilitate the decision of when to conduct a performance peer review:

- Social considerations: Especially the impact on public health, safety, and the environment; impact on national or civil defense; high public visibility of the project; political or social controversy about the project; restriction of normal competition among constructors who can implement the design.
- Technical considerations: Requirement of state-of-the-art expertise; redundancy or "fail-safe" characteristics; extra coordination of a diverse design-construction team; requirement of special construction sequencing.
- Liability considerations: Liability history of the project type; little or no performance history of a unique project; questionable project insurability; project design time that is short or overlapping with the planning or construction phases; existence of critical elements relative to switchover, transfer of loads, transition stages, etc.

21.4 ELEMENTS OF PEER REVIEW

21.4.1 Request for Peer Review

An organizational peer review may be requested in a variety of ways, such as: sought voluntarily by the management of an organization, mandated by an owner as a prerequisite for an approval process, or desired by a large organization for its various operating units.

A project peer review may be requested by the owner, design professional, constructor, or by a government agency that approves the project. The authorization and cost of project peer review is the responsibility of the owner, the party who benefits most from the review.

21.4.2 Scope of Peer Review

An organizational peer review may focus on general procedures for executing projects (as opposed to including all office operations), or it may deal with the complete administrative process, starting with the contract or authorization and proceeding through the programming or preliminary studies, the planning of design phases, the maintaining of records of decisions and other files, the development of basic personal skills, the acquisition of technical equipment and training in its use, the adoption of standards and their distribution, or the teaching of their use. The review might include office facilities in general, libraries, support for field services, and use of "low-tech" or nontechnical machines. Closely related are the policies and practices of "career-pathing" of employees through formal continuing education or on-the-job training. Procedures for sharing experiences might be reviewed as part of the organization's preparation for production. Ultimately, the scope can be broadened to include all facets of an organization's practice, including "staff" functions as well as "line" activities.

The scope of project peer reviews can vary greatly. They may be limited to a review of the final project design documents, or, alternatively, they may be more beneficial and comprehensive if undertaken at defined stages during the design and/or construction process. The scope could be a review of the project plan at the beginning of a project to provide the owner with independent assurance of a well-planned process, or it could include reviews of construction means and methods. There may be only one report at the completion of the review, or there may be progress reports at intermediate stages. The reports may constitute the end of the process, or steps in the process, and evaluation of responses may be included in the peer-review team's charge.

21.4.3 Selection of Reviewers

The peer-review team is independent of the office or project being reviewed. Usually, the reviewers are from outside the organization. However, within the guidelines established earlier in this chapter, they may be from within the organization, but only if they are sufficiently removed from the entity being reviewed.

The review team members selected by the commissioning authority should be well respected by the parties being reviewed. The review team's effectiveness is greatly influenced by the independence, experience, and stature of its members. It is important that the team be composed of senior professionals with sound judgment who are:

- Experienced on similar projects.
- Familiar with governing regulations.
- Objective and thorough.
- Good listeners and skillful communicators.

The size of the team depends upon the scope and complexity of the peer review. A team normally consists of two or more reviewers, selected to cover the various technical disciplines involved. However, less complex projects or organizations could utilize one reviewer.

21.4.4 Review of Documents and On-site Interviews

A peer review begins with a study of written documentation which is generally acquired before the team members are assembled on-site. This is then followed by on-site confidential interviews with a cross section of employees or team members.

In an organizational peer review these interviews provide an opportunity to analyze such elements as organization and project goals, administration, quality assurance/quality control systems, user satisfaction, project control, field supervision and overall direction, and to recommend appropriate changes where needed.

Project peer review consists of examining the information acquired, evaluating its relevance, thoroughness, and accuracy, and drawing conclusions about the status, technical quality, or construction performance of the project from these evaluations. Combining the results of this study with information obtained in on-site interviews with members of the project organization's management and staff, the team can determine the extent to which the design assumptions and project requirements are understood and being followed. The team can also evaluate administrative and technical solutions to problems encountered, and identify innovative features of the project and potential trouble spots.

21.4.5 Reports

An organizational review generally ends with a confidential oral report on the findings of the review to the chief of the reviewed organization. Records of the reviews are typically confidential and may be destroyed after the process. Confidentiality must be maintained because the conduct of an organizational peer review requires the solicitation of personal opinions and comments from all levels of employees who must feel open and free to bring forth their concerns.

The type and distribution of the report must be clearly defined at the start. If an organization is large, with several offices, the results of a peer review of one of the branch offices may be reported to the top management of the organization, as the commissioning authority. Alternatively, an organizational peer review may result in a report only to the office reviewed, for the purpose of its self-improvement.

It is critical that the project peer review team conduct its work in an objective and constructive manner, raising issues and technical questions that will enhance the quality of performance. The end result may be specific recommendations or the identification of areas for further analysis by the project team members.

At the completion of its assigned task, the peer review team prepares a final report covering its activities. This report presents the conclusions of the team, including identification of some areas needing further review. In the case of project-management review, the report could focus on areas that may not have met the terms of the specific project requirements, contracts, or authorizations, and may also provide insight to the organization being reviewed for improvements in the future. In the case of a project-performance peer review, the elements of principal focus could include the suitability of the project design for the owner's needs, the constructor's methods, deficiencies that might affect health and safety, or causes of financial loss to the owner. Of course, all reports should include appropriate favorable comments as well as those necessary for project enhancement.

Reports of project peer reviews of all types are submitted at the times defined in the scope of the peer review. Project-management reports may be oral or written, and, if the latter, they may be in a formal or an outline format. Reports on project management usually do not dictate required actions but are used by management to guide its own response, which is why relative informality and oral reports may be preferable. Design or construction performance reports, on the other hand, should be written with a degree of detail that assures adequate communication of the technical issues that need attention. These technical reports must convey the peer review team's conclusions accurately and thoroughly if they are to be of value.

Reports might be organized as follows:

1. Scope of the review, including its limitations.
2. Status of project.
3. Phase being reviewed.
4. Time period.
5. Items requiring further evaluation and consideration.
6. Recommended corrective actions.

21.4.6 Subsequent Actions

The report of a peer review will presumably note some existing deficiencies and include recommendations for improvement. Something should be expected to happen as a result. In some cases, the commissioning agency will have the authority to order the weaknesses corrected. In other cases, the reviewee has the responsibility to correct its own weaknesses.

Experience has shown that offices which voluntarily seek a peer review (as opposed to having it imposed by some higher authority) are very likely to take the findings seriously and to implement them in the same spirit that led them to request the review in the first place. Such voluntary action can be more effective and, in the long run, more productive than enforced compliance with a program that may not enjoy complete acceptance by the reviewee.

The peer review should have a recognizable end. Usually, the commissioning authority acknowledges that the review has been completed and then releases the review team from further efforts and responsibilities. Certain administrative closeout details, such as financial arrangements, certificates of completion, and the updating of records, should be promptly completed.

21.5 RESPONSIBILITIES OF PARTIES

While the project peer review process is intended to enhance the quality of the constructed project with the input and advice of a second party, it is important to note that the responsibility for project design and construction remains with the project team.

21.6 RECOGNIZED PEER-REVIEW PROGRAMS

There are several types of established organizational peer-review processes. It is essential that any discussion of an organizational peer-review program briefly define the critical parameters of scope, policies, and procedures, budgets, reviewers, frequency of reviews, reports, confidentiality, and the consequences. If an organizational peer review is attempted without these elements outlined in the commissioning document, confusion as to its conduct, meaning, and value will result.

Three peer review programs for design offices are currently available nationally. They are offered by the Association of Engineering Firms Practicing in the Geosciences (ASFE), the American Consulting Engineers Council (ACEC), and the American Society of Civil Engineers (ASCE). The programs have several common elements, but they also have differences. There is also a local peer review program sponsored by the Chicago Chapter of the American Institute of Architects (AIA).

A number of the larger owners, designers, and constructors have established in-house peer-review programs. In these programs, teams of senior officers from different and distant branch offices periodically and methodically review other branches. Generally the organization publishes its policies and procedures for guidance, and the programs generally operate under the administration of the firm's quality management officers who receive strong support at the highest levels.

Several large governmental agencies have begun exploring internal organizational peer reviews also. Presumably, these would be similar to the successful programs established by private design organizations.

CONCLUSION

A peer review is a high-level action taken to improve quality in constructed projects. An organizational peer review examines the policies and practices of a design office across many projects and activities. A project peer review focuses intensely on a single project, perhaps even on a single phase at a time.

Peer reviews are requested as added safeguards for the public, the owner, design professional, and constructor. All groups familiar with peer reviews have encouraged their use by large or small organizations and on large or small projects. A fresh, unbiased, and diplomatic review by one or more independent, high-level professionals can be a highly cost-effective measure and often a means of reducing the overall time required to complete a constructed project.

CHAPTER 22

RISKS, LIABILITIES, CONFLICTS

INTRODUCTION

During the planning, design, and construction of the project, risks to all participants are present. These risks can be broadly classified as: Risk of damage to public health and safety; risk of personal injury to project workers or others; risk of financial loss for a variety of reasons; risk of professional standing or business reputation; risk of litigation among project team members, or third-party litigation.

Risk management can be effected by competent performance of each participant, attention to detail in producing the completed project, and observance of the job-site safety program. Mitigation of financial damage is provided through the use of bonds, warranties, and insurance. Conflicts are best resolved rather than arbitrated or litigated.

Keys to quality in the constructed project include clear, concise contractual arrangements which fairly and realistically align interests of the team members toward common goals; communication which keeps all team members informed and which assists in early identification of problems; teamwork in problem solving and other activities necessary to meet project requirements.

22.1 PROJECT RISKS

Risks to members of the project team resulting from the characteristics of the construction process fall into a number of categories:

- Danger: Construction work has a high potential for job-site injuries to workers and others, including the public.
- Difficulty: Construction work is done under field conditions, using designs that frequently incorporate unique or new ideas, and by a work force with varying degrees of training, skill, and experience.
- Unforeseen Conditions: During the construction phase of the project, previously unknown, unforeseen, or changed conditions may be encountered, requiring modification of design and planned construction activities with attendant cost increases to all parties.
- Diversity of Interests: The owner, design professional, and constructor are in a position of diverse interests involving the allocation of project costs and profits.
- Control: Factors beyond the control of any of the project participants include weather, flood, fire and earthquake, strikes and civil disorders, market conditions affecting availability of materials, actions of regulatory agencies, and others.

Risks in the construction project are managed by the project team members through actions to prevent damage to life and property. If events do occur, the attendant financial damage to team members is mitigated through the use of insurance, bonds, and warranties.

"An Owner's Guide to Saving Money by Risk Allocation," a recent publication by ACEC and AGC, presents a more detailed discussion of risk management.

22.2 PERFORMANCE OF PROJECT TEAM MEMBERS

Risks are managed by the careful performance of each team member in discharging the obligations of the owner/design professional agreement and the owner/constructor contract and in working together as a team to produce a facility meeting project requirements.

22.2.1 Qualifying Project

Before a project is undertaken by the owner, design professional, or constructor, questions relating to feasibility and risk should be evaluated by each team member. Some of the questions which affect feasibility and risk are:

- Will the project meet the owner's needs?

- Is the project adequately funded?
- Is the project schedule realistic?
- What is the project's potential to cause financial loss to the team members?
- What is the project's potential to cause property damage or personal injury?
- What is the project's potential to cause uninsurable losses, e.g., damage from pollution or asbestos contamination?
- Is the project environmentally sensitive?
- Does the project require routine or unique approaches to design and construction?
- Does participation in the project carry the risk of injury to the professional or business reputation of the participants?

If consideration of these questions and others confirms feasibility, and if the risks involved are considered acceptable, the next step is qualifying the team members.

22.2.2 Qualifying Team Members

The success of the effort and quality in the project depend on team members working toward a common goal. Therefore, each team member has a legitimate interest in the attitude, ability, and record of performance of the other team members, particularly as they affect risk. Some of the characteristics are:

- Record of integrity and honesty in business relationships.
- Financial strength and funding capability.
- Record of performance on similar projects.
- Capability and experience of personnel assigned to the project.
- Previous relationship among team members.
- Response to changed conditions—change-order negotiations and acceptance.
- History of litigation on previous projects.

For an individual team member, the qualification of the project and of the other members of the team is a significant element of risk management.

22.2.3 Contractual Arrangements

In structuring the contracts for the project involving the owner, design professional, and constructor, each party may wish to use the services of experienced construction attorneys and insurance advisers.

In the owner/design professional agreement, provisions concerned with risk management include:

- A well-defined scope of services to be provided by the design professional with a statement of what is to be done and what is not to be done.
- A statement of actions, information, or services required of the owner.
- Indemnification for inappropriate risks.
- Limitation of liability when appropriate.
- Waiver of subrogation by insurers.
- Disclaimers of responsibility for certain activities, e.g., constructor's safety programs; design of temporary structures for construction use; means, methods, or sequencing of construction.
- Insurance requirements.
- Conflict-resolution provisions.

In the owner/constructor contract, provisions concerned with risk management include:

- A well-defined scope of work including clear plans, specifications, and other contract documents.
- Indemnification clauses providing protection for the owner and the design professional.
- Limitation of liability when appropriate.
- Responsibility for job-site safety programs and design of temporary structures.
- Waiver of subrogation by insurers when appropriate.
- Quality control programs.
- Insurance and bonds.
- Warranties and guarantees.
- Conflict resolution provisions.

22.2.4 Performance Under Contract

The best protection against the risks involved in the planning, design, and construction of the project is the faithful performance of the contract terms agreed to by the team members. Contracts, however, should not be used to transfer losses from risks to parties that do not have the authority or expertise to control those risks. Attention to detail in completing assignments on time results in quality in the project and reduces the probability of conflict among team members.

The job-site safety program formulated and administered by the constructor for the protection of all participants in the project as well as the general public is an essential tool in reducing the risk of injury to workers and others.

Maintaining lines of communication among the project participants, especially during times of adversity and conflict, helps resolve misunderstandings, define problems, and effect solutions in an atmosphere which encourages teamwork in realizing the common goals.

2.3 BONDS, WARRANTIES, AND INSURANCE

In cases where parties do not perform as required under contractual terms, or when faulty equipment or materials are supplied, or where accidents and natural catastrophes occur, bonds, warranties, and insurance are vehicles provided to mitigate the financial impact of these occurrences.

22.3.1 Bonds

Bonds are used in the project bidding and construction phase generally to protect the owner if the constructor fails to fulfill obligations under the contract or under bidding procedures. Bonds used include:

- Bid bonds: Unless the constructor is able to withdraw a bid because of a mistake, the bid bond will compensate the owner for the difference between the low and the second low bids up to the penal sum of the bond if the bidder does not enter into a contract upon award.
- Performance bonds: The performance bond basically guarantees the creditworthiness of the constructor and provides the owner with financial protection should the constructor default in performance.
- Labor and material payment bonds: The labor and material payment bond guarantees that legitimate bills for labor and materials are paid by the constructor and are used as a means to reduce the risk of liens being placed on the project.

22.3.2 Warranties

Warranties are issued for the purpose of guaranteeing materials and workmanship of the constructor and also for manufactured products incorporated into the project.

- Clauses under the construction contract may require that the constructor return to the project to correct defects in materials and workmanship for a specified period of time after project completion.
- Manufacturers of material may warrant or guarantee their products against defects in workmanship and materials for a specified period.

22.3.3 Insurance

In the course of doing business before, during, and after participation in the project, all team members probably carry insurance for general liability and property damage, motor vehicle operation, and workers' compensation. The risk of project activities imposes additional and special insurance requirements on all parties. Both the owner/design professional agreement and the owner/constructor contract may require the parties to obtain insurance for their own protection, for the protection of other participants, and for the general protection of the project.

22.3.3.1 Insurance Needs of Owner

Owners need insurance coverage for risks imposed by project activities. For this purpose they may purchase, or may require the constructor to purchase, builder's risk coverage, which may include business interruption coverage. The construction contract documents may require that the insurance include interests of the owner, design professional, constructor, and other participants. These requirements are specified by contract and may include policy limits as well as special situations or risks to be covered.

22.3.3.2 Insurance Needs of Design Professional

Design professionals need insurance protection for risks imposed by their project activities which may require that provisions of their usual insurance program be modified to include unique or unusual hazards existing in project design, construction, and operation.

Professional liability (errors and omissions) insurance policies in force may need to be modified to meet special coverages and amounts dictated by characteristics of the project or specified in the owner/design professional agreement. The design professional, in dealing with consultants and others providing services to the project, may impose insurance requirements (professional liability as well as general liability) as a condition of service contracts.

Design professionals, in offering their design for construction and in performing services on the construction site, are placed at risk in situations over which they have little or no control, i.e., injuries to workers, collapse of temporary structures, etc. Insurance protection in these situations is afforded by the design professional's being named as an additional insured on builder's risk and other liability policies carried by the constructor and/or owner for project activities.

22.3.3.3 Insurance Needs of Constructor

In addition to conforming their usual insurance programs to meet specific or unique risks attendant to the project, constructors are usually required to provide insurance coverage under the owner/constructor contract.

Because the constructor supervises the job site, and the construction process puts both life and property at risk, the constructor is usually required to hold harmless and indemnify other parties not in control at the job site. The contract usually stipulates that the constructor support this commitment with insurance coverage, and policy limits are tailored to the size and nature of the project and specified in the contract. To assure that insurance requirements are met, certificates of insurance (or copies of the policy) are filed with the owner, with the provision that insurers must notify the owner if coverage is changed or canceled.

In dealing with subcontractors, suppliers, and manufacturers, the constructor usually requires evidence of insurance or other financial resources. The constructor may elect to extend the coverage of the builder's risk policy to those project participants responsible to the constructor.

22.4 CONFLICT AVOIDANCE

In the broad view, conflicts among the team members can be avoided if attitudes fostering teamwork, open communications, honesty, integrity, and commitment to project goals are present. Actions which help to produce this climate and reduce conflict are:

- Select team members, including the owner, who are professionally and financially capable of performing responsibly.
- Formulate project requirements meeting the needs of all team members, and align interests of team members in matters of time, money, decision making, and performance.
- Structure contracts to define clearly scope of the work and services to be provided, establish lines of authority, and assign responsibilities for owner, design professional, and constructor.
- Perform under contract provisions with competence and on time, and keep appropriate records.
- When unforeseen or changed conditions arise, define the problem, assign new responsibilities to individual team members, and negotiate appropriate payment for extra work.
- Work cooperatively and with flexibility to meet project requirements.

Quality in the constructed project generally results when each team member meets his or her own responsibilities and works harmoniously with the other team members to complete the project.

22.5 CONFLICT RESOLUTION

If a conflict among team members does occur and cannot be resolved by negotiation between the principals it is necessary to move to a more formal means of conflict resolution. Project contracts may have dispute-clause provisions calling for special dispute-resolution processes. Among those processes are:

- Use of the design professional in guiding the owner and constructor to a mutually satisfactory settlement.
- On public contracts, submittal of the controversy to hearing officers or boards of appeals.
- Use of a skilled mediator to help participants structure negotiations and to guide and encourage the participants to reach a settlement.
- Use of a pre-selected contract dispute resolution board.
- Other ADR processes, such as mediation-arbitration; mediation-then-arbitration; neutral expert fact finding and

analysis (binding or nonbinding); "rent-a-judge," and mini-trials.

- Arbitration of the dispute under rules and procedures adopted by the National Construction Industry Arbitration Committee (part of the American Arbitration Association), under procedures specified by contract, or under procedures tailor-made by and mutually acceptable to disputants.

22.6 LITIGATION AS LAST RESORT

Litigation of construction disputes is usually a complex matter involving technical questions relating to planning and design; means, methods, and sequencing of construction; safety programs; interpretation of contracts, and other matters. Before entering into litigation to settle disputes, each participant is well advised to make a dispassionate, objective analysis of the direct and indirect costs of pursuing solution through the courts.

These costs may be classified as follows:

- Direct outside costs: Attorney fees involved in preparing for and trying the case, court costs, deposition costs, expert testimony, special investigations (soils, geology, hydrology, materials).
- Direct inside costs: Cost of personnel working with attorneys in preparation and trial activities, record searches, preparation for depositions and court testimony, attending discovery and trial proceedings, special investigations, and other incidental costs.
- Indirect costs: Interruption of regular duties of key management and technical personnel, reduction of bonding capacity for the constructor, increase in professional liability insurance costs for the design professional, impact on financial standing, interruption of cash flow, impact on professional and business reputation, and other costs.
- Court decision: The court decision may result in an award which more than offsets the accumulated costs—or, it may result in an unfavorable decision which further increases the costs of the lawsuit. In analyzing these possibilities, particular attention should be given to the impact of a downside result.

A thorough analysis of costs and benefits outlined above by individuals not emotionally involved in the dispute is a useful tool in making the decision to enter into litigation. If the analysis is continuously updated it will serve as a guide in settlement discussions as they occur during the progress of the litigation.

CONCLUSION

Risks to public health and safety, as well as the risk of bodily injury, property damage, financial loss, and legal liability can be managed through careful selection of projects and qualification of project team members. Assembling qualified team members capable of working together to meet project requirements and bound by contracts defining the responsibilities of each team member is primary in the management of risk and the achievement of quality in the project.

Insurance, bonds, and warranties afford mitigation of financial damage to project participants or others at hazard during project design and construction. Contracts among the team members specify special forms and limits of these financial vehicles required for protection of the owner, design professional, and constructor. In addition to insurance required by contract, each team member usually carries general liability, workers' compensation, and motor vehicle insurance as a prudent business practice.

The use of insurance to mitigate financial damage does not preclude disputes among project participants as to responsibility for and added costs claimed incidental to the dispute. Conflicts can be avoided by fostering a climate of teamwork among qualified participants working together to meet project requirements. Conflicts not avoided are best resolved by negotiation, mediation, or arbitration. Before embarking on binding arbitration or litigation, each participant is well advised to analyze thoroughly the projected upside and downside results of such actions.

GLOSSARY

ADVERTISEMENT FOR BIDS: Published public notice soliciting bids for a construction project or designated portion of a project, and included as part of the bidding documents.

ALTERNATIVE ANALYSIS: Analysis of the various alternatives to determine validity and impact on project cost, project appearance, project schedule, and socioeconomic and environmental conditions.

ARBITRATION: A method of settling claims or disputes between parties to a contract, used as an alternative to litigation, under which an arbitrator or a panel of arbitrators, selected for specific knowledge in the field in question, hears the evidence and renders a decision.

ARCHITECT: See *Design Professional*.

ARCHITECT-ENGINEER: See *Design Professional*.

BID: A complete and properly signed proposal to perform the construction required by the contract documents, or designated portion of the documents, for an amount or amounts stipulated in the documents. A bid is submitted in accordance with the bidding documents.

BIDDING DOCUMENTS: The advertisement for bids, the instruction to bidders, the bid form, other sample bidding and contract forms, and the contract documents, including any addenda issued prior to receipt of bids.

BID BOND: A form of bid security executed by the bidder as principal and by a surety to protect the owner if the low bidder does not accept the award of contract.

BID FORM: A form furnished to a bidder to be completed, signed, and submitted as the bidder's bid.

BID OPENING: The opening and tabulation of bids submitted before the prescribed bid opening time and in conformity with the prescribed procedures.

BID SECURITY: The deposit of cash, certified check, cashier's check, bank draft, stocks/bonds, money order, or bid bond submitted with a bid. See *Bid Bond*.

BIDDER QUALIFICATION DATA: Information sometimes required by the owner, and sometimes required by law, about the bidder's financial and physical capability to perform the completed construction required by the contract documents, or designated portion of the documents.

BONUS CLAUSE: A provision in the construction contract for payment of a bonus to the constructor for completing the work prior to a stipulated date.

CERTIFICATE OF COMPLETION: A statement prepared by the responsible design professional on the basis of an inspection stating that the work, or a designated portion of the work is, to the best of his or her knowledge, substantially complete.

CHANGE ORDER: A written order to the constructor signed by the owner and/or by the owner's agent or representative, issued after execution of a contract, authorizing a change in the work or an adjustment in the contract sum or the contract time.

CODES: Regulations, ordinances, or statutory requirements of, or meant for adoption by, governmental units relating to building construction and occupancy, adopted and administered for the protection of the public health, safety, and welfare.

CODES OF ETHICS: Official statements prepared by organizations representing members of a profession that establish fundamental principles, canons, and guidelines of practice for the members of that profession.

COMPETITIVE BIDDING: A method, often mandated by law, of selecting constructors for construction projects by price competition between qualified bidders subjected to various rules and procedures.

CONDITIONS OF THE CONTRACT: Those portions of the contract documents that define the rights and responsibilities of the contracting parties and others involved in construction.

CONSTRUCTABILITY ANALYSIS: A review of the ability to construct a project, covering economics, availability of materials, site restrictions, and local conditions that may affect the construction process.

CONSTRUCTION CONTRACT: The agreement or contract between the owner and constructor for construction of a project, or portions of a project, in accordance with contract documents.

CONSTRUCTION MANAGEMENT: Management services provided to an owner by an individual or entity possessing requisite training and experience during the design and/or construction phases of a project.

CONSTRUCTION SUPERVISOR: The constructor's representative at the site who is responsible for continuous field supervision, coordination, and completion of construction.

CONSTRUCTOR: The individual or entity reponsible for performing and completing the construction of a project required by the contract documents. A constructor has a direct contract with the owner.

CONSULTANT: The person or entity who provides specialized advice or services to an owner, design professional, or constructor.

CONTRACT DOCUMENTS: The owner/constructor agreement, the Conditions of the Contract (General, Supplementary, and other Conditions), drawings, specifications, and all addenda issued prior to and all modifications issued after execution of the contract, and any other items that may be specifically stipulated as being included.

COST-BENEFIT RATIO: A comparative procedure that provides a ratio of cost expended to benefits received in terms of present worth.

DESIGN-BUILD: A form of contracting where the constructor is responsible for the design and construction of a facility.

DESIGN DISCIPLINE: A specific category of related professional services such as structural engineering, architecture, mechanical engineering, civil engineering, etc.

DESIGN PROFESSIONAL: A designation reserved, usually by law, for a person or organization professionally qualified and duly licensed to perform architectural or engineering services; which may include but not necessarily be limited to development of project requirements; creation and development of project design; preparation of drawings, specifications, and bidding requirements; and providing of services during the construction and start-up phases of the project.

DESIGN TEAM: The group of individuals or entities representing all design disciplines required for execution of a design contract.

DESIGN TEAM LEADER: The individual responsible for the coordination of design activities of a project. The design team leader also is responsible for monitoring progress and reporting to the owner.

DEVELOPER: An individual or group that arranges for financing and building of a project. Often, the developer is also the owner.

DRAWINGS: Graphic and pictorial documents showing the design, location, and dimensions of the elements of a project.

EJCDC CONSTRUCTION DOCUMENTS: Sample agreements and contracts prepared by the Engineers' Joint Contract Documents Committee (EJCDC).

ENGINEER: See *Design Professional*.

ENGINEER-ARCHITECT: See *Design Professional*.

ENGINEER OF RECORD (EOR): The prime professional engineer or organization legally responsible for the engineering design.

ENVIRONMENTAL IMPACT STATEMENT: (1) A report on the anticipated impact of a proposed project on surrounding conditions. Environmental, engineering, esthetic, and economic aspects are included. (2) A detailed document meeting the goals of the National Environmental Policy Act, discussing alternatives to avoid or minimize adverse impacts or enhance the quality of human environment.

FAST-TRACK CONSTRUCTION: A process whereby design and construction are performed simultaneously. As design is completed for a portion of a project, construction work commences and proceeds for that particular phase. Thus site work will commence as soon as drawings and specifications are available, while, at the same time, structural, mechanical, and electrical design is under way.

INDEMNIFICATION: A collateral contract or assurance by which an individual agrees to secure another against an unanticipated loss or to prevent the other individual from being damaged by the legal consequences of an act or forbearance on the part of one of the parties or of some third party.

INSTRUCTION TO BIDDERS: Instructions contained in the bidding documents for preparing and submitting bids for a construction project or designated portion of a project.

INVITATION TO BIDDERS: See *Advertisement for Bids*.

LIFE-CYCLE COST: The total cost of developing, owning, operating, and maintaining a constructed project for its economic life, including its fuel and energy costs.

LIQUIDATED DAMAGES: A sum established in a construction contract, usually as a fixed sum per day, as the measure of damages incurred by the owner due to the failure of the constructor to complete the work on schedule.

LOSS PREVENTION: The use of safety programs and insurance to mitigate financial losses resulting from loss of life and personal injuries on a construction project.

OBSERVATION: A function of a design professional, requiring visits to a construction site, to become generally familiar with the progress and quality of work and to determine broadly if the work is proceeding according to the contract documents.

OFFICE PRACTICE: A standardized program for a design or construction firm, covering general management of the firm, organization for projects, owner relationships, office procedures, filing and storing material and operational rules.

OWNER: The individual or group that initiates a construction project and is responsible for financing the project.

PLANS: See *Drawings*.

PRE-BID CONFERENCE: A conference arranged by the owner and attended by the design professionals and prospective constructors prior to submission of bids for the purpose of explaining the project, the design professional's intent, and the owner's requirements and expectations, and for responding to questions by the bidders.

PRECONSTRUCTION MEETING: A meeting arranged by the owner after contract award and prior to construction, for the design professional, constructor, and subcontractors to facilitate the initiation of construction operations.

PRESENT-WORTH ANALYSIS: An analysis of costs of a project over an evaluation period, recognizing that money has value with time. The results produce an appropriate method of alternatives comparison.

PROJECT: The total construction as defined by the contract documents.

PROJECT COST: The total cost of the project.

PROJECT EVALUATION: A critical evaluation of a project by the project team members during both design and construction to assess design, schedule, objectives, costs, legal ramifications, and trends that impact cost, quality, and schedule.

PROJECT MANAGEMENT: The planning, organizing, staffing, directing, controlling, and coordination of design and construction activities for a construction project under the direction of a single project manager, who has direct responsibility to and represents the owner.

PROJECT MANAGER: The person or organization, representing the owner, responsible for overall coordination and management of the project activities. The manager may be a member of the owner's, design professional's, or constructor's staff, or an independent contractor employed by the owner.

PROJECT PLAN: A work activity diagram and other documents depicting features of a project's requirements.

PROJECT SCHEDULE: A diagram, graph, or written listing showing proposed and actual times of starting and completion of various elements of design or construction.

PROJECT TEAM: The entities primarily responsible for completing a constructed project: the owner, design professional, and constructor.

QUALIFICATIONS: The attributes of an individual or entity that are judged or reviewed to determine if they meet predetermined standards and requirements.

QUALITY: Conformance to predetermined requirements.

QUALITY ASSURANCE: All planned and systematic actions necessary to provide confidence that items are designed and constructed in accordance with applicable standards and as specified by contract.

QUALITY CONTROL: The examination of services provided and work done, together with management and documentation necessary to demonstrate that these services and work meet contractual requirements.

REASONABLE CARE: A degree of care, precaution, or diligence as may fairly and properly be expected or required, having regard to the nature of the action, or of the subject matter and the surrounding circumstances of the action.

RECORD DOCUMENTS: A compilation of all drawings, specifications, addenda, written amendments, change orders, work directive changes, field orders, and written interpretations and clarifications, maintained in good order and annotated to show all changes made during construction. These record documents, together with all approved samples and a counterpart of all approved shop drawings, are available to the engineer for reference. Upon completion of the work, these record documents, samples, and shop drawings are delivered to the engineer for the owner.

RECORDS: A written account documenting data, activities, transactions, and memoranda of oral communications, and usually including the contract documents.

RESIDENT PROJECT REPRESENTATIVE (RPR): The individual representing the owner, sometimes selected from the design professional's firm, who administers the construction contract and monitors progress and relationships among the project site personnel.

RESPONSIVE BID: A bid meeting the requirements of the bidding documents by a qualified bidder.

RETAINAGE: A sum withheld from progress payments to the design professional or constructor according to terms of owner/designer or owner/constructor agreements.

RISK TRANSFER: Contractual clauses that transfer the risk of project team members to other parties by means of bonds or insurance.

SELECTION COMMITTEE: A committee established by the owner and guided by pre-established criteria and administrative policy, comprised of qualified individuals, including professionals, to make recommendations on selection of design professionals after conducting necessary investigations, interviews, and inquiries.

SHOP DRAWINGS: Drawings, diagrams, schedules, and other data required for manufacture, fabrication, and erection of components of construction and which are prepared by the constructor, subcontractor, or manufacturer.

SPECIAL CONDITIONS: A section of the Conditions of the Contract, other than General Conditions and Supplementary Conditions, which may be prepared to describe conditions unique to a particular project.

SPECIFICATIONS: A part of the contract documents, contained in the project manual, consisting of written requirements for materials, equipment, construction systems, standards and workmanship, and usually including the Conditions of the Contract.

START-UP: Preparing the project or facility for occupancy or use and testing the systems in that facility for operation.

SUBCONTRACTOR: A person or organization who has a direct contract with the constructor.

SUBSTANTIAL COMPLETION: The point where, in the opinion of the engineer, the work is sufficiently complete, in accordance with the contract documents, so that the work (or specified part) can be utilized for the purposes for which it is intended.

SUPPLEMENTARY CONDITIONS: A part of the contract documents which supplements and may also modify, change, add to, or delete from provisions of the General Conditions.

SUPPLIER: A person or firm who supplies materials or equipment for construction, including materials fabricated for a special design.

TASK MANAGER: An individual who manages a specific assignment of the design.

TRADITIONAL SYSTEM: A contractual arrangement (or project-delivery system) for construction consisting of the owner, design professional, and constructor.

TURNKEY: An extension of the design-build contractual arrangement which may include provision for the constructor to provide land acquisition, financing, purchase and installation of industrial process equipment, and other services not included in a standard design-build contract.

UNBALANCED BID: A bid where some of the unit prices do not reflect the actual estimated cost of the services or materials being provided; and the costs of other unit prices are overstated.

VALUE ENGINEERING: An organized approach to identify efficiently costs that do not contribute measurably to a project's quality, utility, durability or appearance, or to the owner's requirements.

ACRONYMS

ACEC: American Consulting Engineers Council

ACI: American Concrete Institute

AGC: Associated General Contractors of America, Inc.

AIA: American Institute of Architects

AISC: American Institute of Steel Construction

ANSI: American National Standards Institute

ASCE: American Society of Civil Engineers

ASFE: The Association of Engineering Firms Practicing in the Geosciences

ASTM: American Society for Testing and Materials

CAD: Computer-aided Design; or Computer-aided Drafting

CMAA: Construction Management Association of America

CPM: Critical Path Method

CSI: Construction Specifications Institute

EEO: Equal Employment Opportunity

EJCDC: Engineers Joint Contract Documents Committee

FAR: Federal Acquisition Regulations

NAS: Network Analysis Systems

NEPA: National Environmental Policy Act

NSPE: National Society of Professional Engineers

O&M: Operations and Maintenance

PCI: Prestressed/Precast Concrete Institute

PDM: Precedence Diagramming Method

QA: Quality Assurance

QC: Quality Control

QA/QC: Quality Assurance/Quality Control

RPR: Resident Project Representative

APPENDIX 1

BRIEF HISTORY OF DEVELOPMENT OF *QUALITY IN THE CONSTRUCTED PROJECT*

In November 1984 the American Society of Civil Engineers (ASCE) sponsored a workshop in Chicago for approximately 100 construction industry leaders on the subject of "quality in the constructed project." At the time, ASCE, other organizations, and industry officials from across the country were concerned about the increasing number of project failures, disappointments, accidents, and other problems that resulted in considerable cost overruns, excessive litigation, injury, and even death. The principal recommendation from that workshop was that ASCE should develop and publish a comprehensive and authoritative guide on quality in design and construction that would clearly define roles and responsibilities of the participants in the process.

In early 1985 ASCE President Richard Karn appointed the following to serve on a steering committee to supervise the Guide's development:

- James W. Poirot, Chairman of the Board of CH2M-Hill Companies Ltd., Denver, CO—Chairman
- John D. Stevenson, President, Stevenson & Associates, Cleveland, OH
- Harold L. Loyd, Senior Vice President, Turner Collie & Braden, Inc., Houston, TX
- Jerome S.B. Iffland, President, Iffland Kavanagh Waterbury, New York, NY
- Robert L. Morris, Executive Vice President, Wildman & Morris, Inc. San Francisco, CA
- R. Lawrence Whipple, ASCE Managing Director of Professional Affairs— Staff representative.

The committee developed a scope, schedule, and budget. With the assistance of a consulting librarian, a subject-matter outline and table of contents were developed from numerous resource materials supplied by owners, designers, and constructors. The committee then selected the authors, reviewers, and format, and expanded on the overall goals and objectives for the project. The broad goals of the Guide, as outlined by the committee, are to provide quality guidelines and exemplary procedures for the primary participants in a constructed project. Specifically, the Guide outlines responsibilities, duties, and limits of authority for owners, designers, and constructors in each phase of the design and construction process. It also covers topics that play a critical role in producing or improving quality in construction, such as contract documents, procedures for selecting design professionals, the role of an owner in constructed projects, the role of the design professional during construction, liability of project-team members, peer review, and many other important aspects of design and construction.

The project currently consists of two volumes. The 22 chapters in Volume 1 cover the entire design and construction process from conception to start-up and, with two exceptions, do not focus on specific disciplines or practices. Volume 1 chapters cover the importance of such aspects as "The Owner's Role, Expectations, and Requirements," "The Project Team," "Procedures for Selecting the Design Professional," and "The Construction Team." The 14 Volume 2 chapters, which are currently under development, do cover specific technical disciplines, among them "Structural Engineering," "Geotechnical Engineering," "Transportation," "Urban Planning and Development," and others.

The committee recruited approximately 40 authors and 70 reviewers to produce Volume 1. They represent a broad spectrum of design and construction and are from every segment of the industry, including private firms, public agencies, universities, trade organizations and publications, and the U.S. government. A managing editor was hired to coordinate their efforts and edit the drafts. A technical editor, Leslie A. Clayton, reviewed the drafts for specific technical accuracy and consistency. Mr. Clayton is a former chairman of Boyle Engineering Corporation in Newport Beach, California.

Early in the development stage the committee decided that, due to the Guide's scope, various industry organizations should have the opportunity to assist in developing its contents. Consequently, a briefing was held in New York City to which representatives of all identified related organizations were invited. The purpose of the briefing, which was attended by 53 representatives of various organizations and publications, was to inform them of the purpose of the project and to invite their participation. Several of these groups recommended authors and reviewers and indicated an interest in being involved.

An initial review draft of Volume 1 was distributed in October 1986. In early April 1987 more than 1,000 copies of a revised draft of Volume 1 were distributed for general review. The editors, authors, and committee members reviewed over 800 pages of comments and made extensive revisions to the text. The draft was also reviewed by an experienced legal counsel, John Clark, former counsel to the Engineers Joint Contract Documents Committee, who submitted 90 pages of comments, the majority of which were incorporated into the text.

In May 1988 a third draft, entitled "A Preliminary Edition for Trial Use and Comment," was released. The trial use and comment period extended until December 1989. During that period owners, designers, constructors, and others involved in the design and construction process were urged to use the Guide, when possible, on their projects, and to submit comments on its contents.

Realizing that implementing trial use of the Preliminary Edition would be a complicated and critical task, the Guide's developers and supporters agreed that a special task committee should be created to coordinate this activity. In the fall of 1987 such a committee was appointed. Its members were:

Stephen C. Mitchell, Executive Vice President, Lester B. Knight & Associates, Inc., Chicago, IL—Chairman
Donald H. Kline, Kimley-Horn & Associates, Raleigh, NC
Robert A. Perreault, Jr., Perreault and Boisvert, Manchester, NH
A. C. Burkhalter, Friendswood Development Co., Houston, TX
Dean E. Stephan, Charles Pankow Builders, Co., Altadena, CA

Under their leadership the Guide was strategically placed among educators; major public owners, including the Corps of Engineers, NAVFAC, and state Departments of Transportation; private sector CEOs; public officials, including the governors of the 50 states and the mayors of 20–30 large cities; and ASCE section and branch presidents. In addition, more than 15,000 copies were purchased by users from all walks of the construction industry.

The Implementation Committee developed materials to maximize the Guide's usefulness to its recipients. Included in their projects were a sample course outline for use at the college level; case study questionnaires; outlines for student chapter/club programs, in-house discussion groups, and section/branch programs; and videotape and slide presentations.

The Guide was the subject of numerous presentations at ASCE section/branch meetings, technical society meetings both in the U.S. and abroad, including meetings in Japan, Korea, Australia, Sweden, and Scotland, as well as at a FIDIC meeting in Washington, D.C.

The Construction Industry Institute conducted a comprehensive review of the Guide. The initial plan was to apply it to the construction of an EXXON packaging plant in Baton Rouge. Scheduling conflicts made this impossible; however, the committee, which included representatives of EXXON, IBM, DuPont, MK/Ferguson, and Zachary Construction, presented a thorough review, much of which is reflected in this first edition.

The American Consulting Engineers Council submitted extensive comments compiled from reviews by many of its state societies. A data base of 1500 ± comments from individuals was compiled, and input from approximately 12 ASCE sections and branches, four overseas engineering organizations, 12 related societies, and three government agencies was also considered.

A panel of attorneys and insurance professionals reviewed the Guide during the summer of 1989 and presented the results of their review in a two-day Legal Forum in September 1989 at Keystone, Colorado. Many of their comments on the Preliminary Edition are reflected in the revised text. The panel members, who collectively represent owners, designers, and constructors, were not employed as legal or insurance advisers to ASCE; their participation was both voluntary and independent of the firms and/or organizations with which they are affiliated. We are indebted to them for their input.

The editors considered hundreds of pages of comments and suggestions in preparing the current edition. As each chapter was revised it was sent to its original authors, the steering committee, the Legal Forum, ASCE counsel, and representatives designated by the presidents of ACEC and NSPE for review. The construction chapters had the added benefit of review by representatives of AGC who produced much useful comment in a very brief turnaround time.

The Steering Committee is deeply grateful for the time, effort, and contributions made by the numerous authors and reviewers. They have performed an invaluable service to their profession and industry. We are also indebted to the hundreds who read the Guide, applied it to their practice, and gave us the benefit of their comments.

ASCE plans to maintain the Guide as a living document. A Committee on Quality in the Civil Engineering Profession has been appointed and charged, in part, with ongoing review and revision of the Guide at regularly scheduled intervals. Readers are urged to participate in this continuing process by addressing their comments to the Manager of Professional Services, ASCE, 345 East 47th Street, New York, New York 10017.

CHAPTER AUTHORS

Glen Abplanalp

Robert R. Adams
CH2M-Hill, Inc.

Thor Anderson
Green & D'Angelo

Robert P. Bayless
George F. Young, Inc.

Edward P. Becker
Consulting Engineer

Mark B. Bodner
Lehrer McGovern Bovis, Inc.

Holly A. Cornell
CH2M-Hill, Inc.

Robert C. Esenwein
Turner Collie & Braden, Inc.

Lee S. Garrett
Turner Collie & Braden, Inc.

Kenneth Gibble
Besier Gibble Norden

Joe Gutierrez
Los Alamos National Laboratory

C. Raymond Hayes
Rust International

Melville D. Hensey
Hensey Associates

Richard E. Holmes
Lorenz & Williams, Inc.

Richard Hovey
REA Consulting Engineers

John P. Hribar
Howard Needles
Tammen & Bergendoff

Dale Kern
Peterson & Co.

Donald H. Kline
Kimley-Horn & Associates, Inc.

Alfred C. Leonard
Malcolm Pirnie, Inc.

Wyatt McCallie
CH2M-Hill, Inc.

Arthur R. Meenen
Sverdrup Corp.

Walter P. Moore
Walter P. Moore & Assocs.

Michael Moshiri
L.A. County Sanitation District

Glenn S. Orenstein
Stone & Webster Engineering Corp.

Charles A. Parthum
Camp Dresser & McKee, Inc.

James H. Pope
Pacific Gas & Electric Co.

Orrin Riley
Orrin Riley Engineers

Robert D. Rowland
Herbert, Rowland & Grubic, Inc.

Howard A. Schirmer, Jr.
CH2M-Hill, Inc.

Robert J. Smith
Wickwire, Gavin

C.B. Tatum
Stanford University

Richard Tomasetti
Thornton-Tomasetti Assocs.

Donald R. Trim
Wade-Trim Group, Inc.

C. Roy Vince
Prof'l Liability Broker & Cslts

Carlton S. Wilder
Camp Dresser & McKee, Inc.

Shelby K. Willis
Bucher Willis & Ratliff

Phillip W. Worrall
Browne, Worrall & Johnson, Inc.

Joseph E. Worth
CH2M-Hill, Inc.

Richard N. Wright
Natl. Inst. of Standards and Technology

Chuck S. Zickefoose
Brown & Caldwell

LEGAL FORUM

Robert A. Rubin, Chairman
Elliott Gleason
Gary Gough
David Hatem
Bryan Hickey
Paul Lurie
Ron Martell
Robert Meyers
Manning Seltzer

APPENDIX 2

REPRESENTATIVE LIST OF STANDARD-FORM AGREEMENTS

Documents, Forms, Commentaries and Other Publications of the ENGINEERS JOINT CONTRACT DOCUMENTS COMMITTEE

1910-1	Standard Form of Agreement Between Owner & Engineer for Professional Services
1910-1A	Suggested Listing of Duties, Responsibilities and Limitations of Authority of Resident Project Representative
1910-2	Standard Form of Letter Agreement Between Owner and Engineer for Professional Services
1910-8	Standard General Conditions of the Construction Contract
1910-8A1	Standard Form of Agreement Between Owner and Contractor on the Basis of a Stipulated Price
1910-8A2	Standard Form of Agreement Between Owner and Contractor on the Basis of Cost-Plus
1910-8B	Change Order
1910-8D	Certificate of Substantial Completion
1910-8E	Application for Payment
1910-8F	Work Directive Change
1910-9	Commentary on Agreements for Engineering Services and Construction Related Documents by John R. Clark, Esq.
1910-9A	Commentary on Construction Related Documents by John R. Clark, Esq.
1910-9B	An Update to Indicate Important Changes in EJCDC Standard Agreements for Engineering Services
1910-9C	Focus on Shop Drawings by John R. Clark, Esq.
1910-9D	Recommended Competitive Bidding Procedures for Construction Projects by Robert J. Smith, Esq.
1910-9E	Limitation of Liability in Design Professional Contracts
1910-9F	Guide Sheet for Including Limitation of Liability in the Standard Form of Agreement Between Owner and Engineer for Professional Services
1910-9G	Indemnification by Engineers—A Warning
1910-10	Standard Form of Agreement Between Engineer and Architect for Professional Services
1910-11	Cross Reference Between EJCDC Standard General Conditions of The Construction Contract and AIA General Conditions of the Contract for Construction.
1910-12	Guide to the Preparation of Instructions to Bidders
1910-13	Standard Form of Agreement Between Engineer and Associate Engineer for Professional Services
1910-14	Standard Form of Agreement Between Engineer and Consultant for Professional Services
1910-16	Uniform Location of Subject Matter
1910-17	Guide to the Preparation of Supplementary Conditions
1910-18	Suggested Bid Form and Commentary for Use

1910-19	Standard Form of Agreement Between Owner and Engineer for Study and Report Professional Services
1910-20	Engineer's Letter to Owner Requesting Instructions re Bonds and Insurance During Construction
1910-21	Owner's Instructions to Engineer re Bonds and Insurance During Construction
1910-22	Notice of Award
1910-23	Notice to Proceed
1910-24	Contract Documents Bibliography
1910-25	Advice to Engineers Who Intend Using the 1987 Editions of the AIA's Documents
1910-26A	Standard Form of Procurement Agreement Between Owner and Contractor
1910-26B	Procurement General Conditions
1910-26C	Guide to the Preparation of Procurement Supplementary Conditions
1910-26D	Instructions to Bidders for Procurement Contracts
1910-26E	Commentary on Procurement Documents by John R. Clark, Esq.
1910-27A	Standard Form of Agreement Between Owner and Geotechnical Engineer for Professional Services
1910-27B	Standard Form of Agreement Between Engineer and Geotechnical Engineer for Professional Services
1910-28A	Construction Performance Bond
1910-28B	Construction Payment Bond

Documents Prepared or Endorsed by the ASSOCIATED GENERAL CONTRACTORS OF AMERICA, INC.

Design-Build Documents

400	Preliminary Design-Build Agreement
405	Design-Build Guidelines
410	Standard Form of Design-Build Agreement and General Conditions Between Owner and Contractor (Provides a Guaranteed Maximum Price)
415	Standard Form of Design-Build Agreement and General Conditions Between Owner and Contractor (Where the Basis of Compensation is a Lump Sum)
420	Standard Form of Agreement Between Contractor and Architect
430	Conditions Between Contractor and Subcontractor for Design-Build
440	Change Order/Contractor Fee Adjustment
450	Standard Design-Build Subcontract Agreement with Subcontractor not Providing Design
450-1	Standard Design-Build Subcontract Agreement with Subcontractor Providing Design

Construction Management Documents

500	Standard Form of Agreement Between Owner and Construction Manager
501	Amendment to Owner-Construction Manager Contract
510	Standard Form of Agreement Between Owner and Construction Manager (Owner Awards All Trade Contracts)
520	General Conditions for Trade Contractors under Construction Management Agreements

525	Change Order/Construction Manager Fee Adjustment
540	Construction Management Guidelines
545	Construction Management Control Process
550	Owner Guidelines for Selection of a Construction Manager
560	Construction Management Delivery Systems for Hospital Facilities
590	Construction Management Kit

Subcontract Documents

600	Standard Subcontract Agreement for Building Construction
601	Subcontract for Use on Federal Construction
603	Short Form Subcontract
605	Standard Subbid Proposal
606	Subcontract Performance Bond
607	Subcontract Payment Bond
610	Subcontractor's Application for Payment
614	Invitation to Bid Form for Subcontractors

Other

201	Contract Documents for Construction of Federally Assisted Water and Sewer Projects
221	Standard Questionnaires and Financial Statement for Bidders (for Engineering Construction)
230	Wastewater Treatment Standard Proposal Form
625	AGC Certificate of Substantial Completion
630	First Budget Estimate Guideline
650	An Owner's Guide to Building Construction Contracting Methods
690	Guidelines for Obtaining Owner Financial Information/Owner Financial Questionnaire

Contract Documents of the AMERICAN INSTITUTE OF ARCHITECTS

A SERIES/Owner-Contractor Documents

A101	Owner-Contractor Agreement Form—Stipulated Sum
A101/CM	Owner-Contractor Agreement Form—Stipulated Sum—Construction Management Edition
A107	Abbreviated Owner-Contractor Agreement Form for Construction Projects of Limited Scopes
A111	Owner-Contractor Agreement Form—Cost Plus Fee
A117	Abbreviated Owner-Contractor Agreement Form—Cost Plus Fee
A171	Owner-Contractor Agreement for Furniture, Furnishings and Equipment
A177	Abbreviated Owner-Contractor Agreement for Furniture, Furnishings and Equipment
A191	Standard Form of Agreements Between Owner and Design-Builder
A201	General Conditions of the Contract for Construction

A201/CM General Conditions of the Contract for Construction—Construction Management Edition
A201/SC Federal Supplementary Conditions of the Contract for Construction
A271 General Conditions of the Contract for Furniture, Furnishings and Equipment
A305 Contractor's Qualification Statement
A310 Bid Bond
A311 Performance Bond and Labor and Material Payment Bond
A311/CM Performance Bond and Labor and Material Payment Bond—Construction Management Edition
A312 Performance Bond and Payment Bond
A401 Contractor-Subcontractor Agreement Form
A491 Standard Form of Agreements Between Design-Builder and Contractor
A501 Recommended Guide for Competitive Bidding Procedures and Contract Awards
A511 Guide for Supplementary Conditions
A511/CM Guide for Supplementary Conditions—Construction Management Edition
A512 Additions to Guides for Supplementary Conditions
A521 Uniform Location Subject Matter
A571 Guide for Interiors Supplementary Conditions
A701 Instructions to Bidders
A771 Instruction to Interiors Bidders

B SERIES/Owner-Architect Documents

B141 Standard Form of Agreement Between Owner and Architect
B141/CM Standard Form of Agreement Between Owner and Architect—Construction Management Edition
B151 Abbreviated Owner-Architect Agreement Form
B161 Standard Form of Agreement Between Owner and Architect for Designated Services
B161/CM Standard Form of Agreement Between Owner and Architect for Designated Services—Construction Management Edition
B162 Scope of Designated Services
B171 Standard Form of Agreement for Interior Design Services
B177 Abbreviated Interior Design Services Agreement
B181 Owner-Architect Agreement for Housing Services
B352 Duties, Responsibilities, and Limitations of Authority of the Architect's Project Representative
B431 Architect's Qualification Statement
B511 Guide for Amendments to AIA Document B141
B512 Amendment to Standard Form of Agreement Between Owner and Architects
B727 Standard Form of Agreement Between Owner and Architect for Special Services
B801 Standard Form of Agreement Between Owner and Construction Manager
B901 Standard Form of Agreements Between Design-Builder and Architect

C SERIES/Architect-Consultant Documents

C141 Standard Form of Agreement Between Architect and Consultant

C142	Abbreviated Form of Agreement Between Architect and Consultant
C161	Standard Form of Agreement Between Architect and Consultant for Designated Services
C727	Standard Form of Agreement Between Architect and Consultant for Special Services
C801	Joint Venture Agreement

G SERIES/Architect's Office and Project Forms

G601	Land Survey Requisition
G602	Geotechnical Services Agreement
G604	Professional Services Supplemental Authorization
G605/606	Purchase Order and Purchase Order Continuation Sheet
G612	Owner's Instructions for Bonds and Insurance
G701	Change Order
G701/CM	Change Order—Construction Management Edition
G702	Application and Certificate for Payment
G702/CR	Continuous Roll for Application and Certificate for Payment
G703	Continuation Sheet for G702
G703/CR	Continuous Roll Continuation Sheet for G702
G704	Certificate of Substantial Completion
G705	Certificate of Insurance
G706	Contractor's Affidavit of Payment of Debts and Claims
G706A	Contractor's Affidavit of Release of Liens
G707	Consent of Surety to Final Payment
G707A	Consent of Surety to Reduction in or Partial Release of Retainage
G709	Proposal Request
G710	Architect's Supplemental Instructions
G711	Architect's Field Report
G712	Shop Drawing and Sample Record
G714	Construction Change Directive
G722	Project Application and Project Certificate for Payment
G723	Project Application Summary
G801	Application for Employment
G804	Register of Bid Documents
G805	List of Subcontractors
G807	Project Directory
G809	Project Data
G810	Transmittal Letter
G811	Employment Record
G813	Temporary Placement

APPENDIX 3

Recommended Competitive Bidding Procedures for Construction Projects

by

ROBERT J. SMITH, P.E., ESQUIRE

Prepared for
ENGINEERS JOINT CONTRACT DOCUMENTS COMMITTEE

and
Issued and Published Jointly By

PROFESSIONAL ENGINEERS IN PRIVATE PRACTICE
A practice division of the
NATIONAL SOCIETY OF PROFESSIONAL ENGINEERS

AMERICAN CONSULTING ENGINEERS COUNCIL

AMERICAN SOCIETY OF CIVIL ENGINEERS

CONSTRUCTION SPECIFICATIONS INSTITUTE

This document has been prepared in cooperation with

The Associated General Contractors of America

EJCDC No. 1910-9-D (1987 Edition)

National Society of Professional Engineers
1420 King Street, Alexandria, VA 22314

American Consulting Engineers Council
1015 15th Street, N.W., Washington, DC 20005

American Society of Civil Engineers
345 East 47th Street, New York, NY 10017

Construction Specifications Institute
601 Madison Street, Alexandria, VA 22314

Table of Contents

INTRODUCTION 1

I. USE OF COMPETITIVE BIDDING 2

Public Work 2
Private Work 2
Maximizing the Advantages of Competitive Bidding 2
Minimizing the Disadvantages of Competitive Bidding 2
Role of the Owner and the Engineer in Selecting and Implementing Competitive Bidding Procedures 3

II. FORMS AND DOCUMENTS USED IN THE BIDDING PROCESS 4

Introduction 4
Advertisement and Invitation to Bid 4
Instructions to Bidders 5
Bid Form 5
Information Available to Bidders 5
Bidder Qualification Data 5
Bid Security/Bid Bonds 5
Addenda 6
Subcontractor Listing 6
Restrictive Specifications 6
Preprinted General Conditions 6

III. RECOMMENDED PROCEDURES—PRIOR TO BID OPENING 7

Selection of Prospective Bidders 7
Number of Bidders 7
Procedures for Advertising for Bids 7
Time for Preparation of Bids 8
Bid Opening Date, Time, and Place 8
Issuance of Bidding Documents 8
Prebid Inquiries 8
Issuance of Addenda 9
Prebid Conferences 9
Site Investigation Requirements 9
Withdrawal of Bid 9
Bid Modification 9

IV. RECOMMENDED PROCEDURES—BID OPENING 10

Late Bids and Late Modifications to Bids 10
Opening Bids 10

V. RECOMMENDED PROCEDURES—AFTER BID OPENING 12

Introduction 12
Bidder Evaluation—Responsibility 12
Bid Evaluation—Responsiveness 13
Mistakes; Correction and Withdrawal of Bids 13
Dealing with an Unusually Low Bid 13
Determining the Lowest Bid 13
Unbalanced Bids 14
Rejection of All Bids—When Permitted 14
Limitations on Negotiating With the Low Bidder 14

VI. RECOMMENDED PROCEDURES—CONTRACT AWARD 16

The Firm Bid Rule 16
Engineer's Recommendation 16
Legal Review 16
Dealing With Protests and Lawsuits 16
Notice of Award 17
Contractor Response to Notice of Award 17
Executed Counterparts 17
Federal and State Requirements 17
Notice to Proceed 17

APPENDIX 18

RECOMMENDED COMPETITIVE BIDDING PROCEDURES FOR CONSTRUCTION PROJECTS

Introduction

Competitive bidding is a widely used method of obtaining and selecting contractors for construction projects. Often mandated by law, the system is subject to an assortment of rules and procedures. The EJCDC believes that if bidding procedures are clear and detailed, all parties to the construction process will benefit. Contractors will be motivated to bid when they can see the ground rules are fair and clearly set forth. Also, there will be a more orderly and less dispute prone bidding phase which in turn will allow everyone to focus on the objective of entering into a contract and constructing a project.

Many competitive bidding procedures are contemplated or specified in some of the EJCDC documents, e.g., Document 1910-12, *Guide to the Preparation of Instructions to Bidders*. Other procedures are not however. These recommendations have been compiled to help fill this void. Of course, applicable legal requirements or long-established owner policies must necessarily supercede these general recommendations. The EJCDC has published these recommendations to serve as a reference to the owner, the engineer, and contractor in planning and participating in the competitive bidding process.

The EJCDC believes that adherence to the procedures outlined herein will facilitate improved competitive bidding, which in turn will result in savings to the owner. The following sections deal with each step in the bidding process sequentially—procedures prior to bid opening, procedures at bid opening, procedures after bid opening, and procedures for contract award.

CHAPTER I

Use of Competitive Bidding

Public Work

Competitive bidding (sometimes called formal advertising or sealed bidding) is the widely used method of contractor selection and contract formation for federal, state and local government construction projects throughout the country. Typically, the use of competitive bidding is mandated by law or regulation. This mandate reflects a legislative body's recognition that competitive bidding provides value to the taxpayers and fairness in placement of major sums of taxpayer-funded work. Beyond this strong general preference, there are a host of specific rules and criteria that vary with the particular public owner involved. It is a reasonable generalization to state that fairness and integrity are paramount goals.

Private Work

While private owners are not required to use competitive bidding, they frequently find it advantageous to do so because of the benefits of price competition. Unlike a public agency, a private owner has broad discretion in dealing with competitive bids and bidders. For example, bid openings may be private, and contract award need not necessarily be made to the low bidder.

Maximizing the Advantages of Competitive Bidding

The principal advantage in competitive bidding is expressed in the name of the system itself—competition. It is universally recognized that for true competition to exist, there must be clear, complete and detailed drawings and specifications that define the desired effort, and that the invitation be widely distributed to encourage a sufficient number of bidders to undertake the effort to prepare and submit a bid. In addition, major risks associated with the project should be identified and allocated in the contract documents.

Minimizing the Disadvantages of Competitive Bidding

Competitive bidding has been criticized because of the emphasis on award to a low bidder and because it sometimes generates fierce disputes over contract award. It is possible, through the proper selection and application of "responsibility" criteria (see pages 12–13), to reduce the risk of contract award to a bidder who lacks the ability to properly perform the contract. Where permitted by law, prequalification of bidders by the owner may be advantageous. Among other advantages, prequalification eliminates bidders whose chances of contract award are remote before such bidders expend time and effort to prepare a bid. The owner's prequalification should consider all factors as discussed under the topic "Bidder Evaluation-Responsibility" in Chapter V of this publication applied to the specific project for which bids are solicited. If there is no prequalification, the project-specific and general factors which will be applied to determine responsibility should be included in the Instructions to Bidders.

Role of the Owner and the Engineer in Selecting and Implementing Competitive Bidding Procedures

Naturally the degree of involvement of the owner and the engineer will vary depending on the owner's professional staff size and capability, and the assignment undertaken by the engineer. A typical scenario would have the engineer carrying out responsibilities such as those specified in EJCDC Document 1910-1, *Standard Form of Agreement Between Owner and Engineer for Professional Services*, which provides in pertinent part:

> 1.4.4. Prepare for review and approval by OWNER, its legal counsel and other advisors contract agreement forms, general conditions and supplementary conditions, and (where appropriate) bid forms, invitations to bid and instructions to bidders, and assist in the preparation of other related documents.
>
> . . .
>
> 1.5.1. Assist OWNER in obtaining bids or negotiating proposals for each separate prime contract for construction, materials, equipment and services.
>
> . . .
>
> 1.5.4. Assist OWNER in evaluating bids or proposals and in assembling and awarding contracts.

It is important that the owner and engineer agree that a specific series of procedures (such as those discussed in this booklet) be followed. Once this is done, the allocation of responsibility can be readily accomplished. The checklist in the Appendix may be useful in doing this.

In the case of preparation of documents or procedures for a public agency competitive bid, the owner's legal counsel should be consulted with respect to mandatory inclusions in the documents as well as limitations on rights that might be expressed in standard documents.

CHAPTER II

Forms and Documents Used in the Bidding Process

Introduction

The Advertisement or Invitation to Bid (sometimes called "Legal Notice"), the Instructions to Bidders, Information for Bidders, and bid forms are "Bidding Requirements", as distinguished from "Contract Documents."[1] The "Bidding Documents" include the "Bidding Requirements" and the "Contract Documents." Owners and engineers preparing Bidding Documents should use Invitations to Bid and Instructions to Bidders only for their intended purposes, i.e. to be procedural and informative but not contain substantive contractual provisions, and should scrupulously avoid any restatement or explanation of technical or contractual provisions covered in the Contract Documents.[2]

Advertisement and Invitation to Bid

The Invitation to Bid is typically a short document summarizing key information about the project. Its purpose is to attract prospective bidders, announce the bidding schedule for the project and give sufficient information for prospective bidders to determine whether they should obtain copies of the bidding documents. A companion item is the text for an Advertisement which may be placed in newspapers of general circulation and appropriate trade publications to further publicize the upcoming project. Good examples of Invitations and Advertisements may be found in the chapter entitled "Bidding Requirements" of the Construction Specifications Institute *Manual of Practice*.

---

[1]This terminology is based on that used in EJCDC documents. See e.g., *Guide to Preparation of Instructions to Bidders*, Doc. 1910-12, *Standard General Conditions for Construction Contract*, EJCDC Doc. 1910-8, and *Uniform Guide to Locating Subject Matter*, Doc. 1910-16. The EJCDC defines the Contract Documents as the initial Owner-Contractor Agreement, Addenda pertaining to the Contract Documents, the Bid (sometimes), Performance and Payment Bonds (if required), General Conditions, Supplementary Conditions, Specifications, and Drawings. Thus, EJCDC documents can be used when a contract is to be negotiated and not bid. Note that the Bidding Requirements (except the Bid Form when identified as a Contract Document) are not operative subsequent to formation of a contract. Thus, it is important to insure that subject matter applicable to the contractor's performance is not included in the Bidding Requirements.

[2]This philosophy is consistent with the *Uniform Guide to Location of Subject Matter*, EJCDC Doc. 1910-16.

Instructions to Bidders

The Instructions to Bidders are the rules with which a bidder must comply during the bidding process. EJCDC's Document 1910-12, *Guide to the Preparation of Instructions to Bidders*, is recommended as a starting point for the preparation of such a document. This guide is intended to be used for bidding when the contract documents contain the EJCDC *General Conditions*, Document 1910-8. Note that EJCDC no longer publishes a "standard form" of instructions, reflecting its view that the instructions should be developed on a project-specific basis. Indeed, it is through the instructions that many bidding procedures will be imposed. Thus, preparation of the instructions for a particular project should be based on decisions made consistent with Chapter III of this publication.

Bid Form

The bid form on which the bidder submits its price quotation is a significant document. When completed and signed, it is legally deemed an offer to enter into a contract on the terms and conditions stated in the bid form.

Engineering projects may be bid on a variety of bases, e.g. unit price, lump sum, and may be subject to varying legal requirements. EJCDC's *Suggested Bid Form and Commentary for Use*, Document 1910-18, is a useful guide for preparing a project-specific bid form. Bids should be required to be submitted in writing on a standard form to ensure uniform arrangement of pricing information. This facilitates comparisons and permits exact comparison of bids.

One copy of the bid form should be bound in the Project Manual and separate bid forms should also be provided with the bidding documents. Although at times it is required, it should normally not be necessary for bidders to complete a bid form bound in a project manual and submit the entire bound manual. There should be only one "original" of the bid—all additional copies should be reproduced from the "original" to avoid mistakes.

On engineered construction, competition is enhanced by an accurate estimate of quantities for unit price items. Additionally, the documents should clearly define the scope of the work and all typical details for each item of work upon which unit price payments will be based.

In the case of unit price contracts the instructions should state that the estimated quantities are approximate only and that the resultant price for the work is an estimated total price in the proposal itself. In determining the low bidder, written unit prices should take precedence over numerals. A specific section of the unit price contract should provide for an equitable adjustment of unit prices in the event of a specified over-run or under-run of estimated quantities.

Information Available to Bidders

Identification of the nature and location of information in the owner's or engineer's possession that would be useful to bidders in preparing their bids is highly recommended. Courts in many states have imposed a duty to disclose such information on both owners and engineers. The EJCDC General Conditions at paragraphs 4.2 and 4.3 contemplate such disclosure. Further details and recommendations for specific language are contained in EJCDC's *Guide to the Preparation of Supplementary Conditions*, Document 1910-17.

Bidder Qualification Data

Some laws require—and some owners desire—that bidders provide certain information relative to their financial and physical capability to perform the work. A blank form of such a document is often included in the bidding documents. Owners, engineers, and other parties must realize this information is extremely confidential. Improper use or disclosure of such information can be damaging to a contractor. Only essential and pertinent information should be required of bidders. More detailed information may be obtained from the apparent low bidder after bid opening. Standard forms are often used for this purpose.[3]

Bid Security/Bid Bonds

A specimen bid bond is often included in the bidding documents, although many actual bid bonds are received on preprinted forms of surety companies.[4] The purpose of such security is to help insure that the low bidder will execute a contract if it is awarded. The requirement for such a bond is often mandated by law. Bid bonds are more widely utilized than certified checks or other negotiable instruments, which are normally permitted. A bid bond normally constitutes a representation that the surety will provide any required performance and payment bonds if the bidder is awarded the contract.

The amount of such security is most commonly five to ten percent of the bid.

Bid bonds are one of two types—actual damages (difference in price) or liquidated damages. The actual damages bond is essentially like a security deposit and

[3] EJCDC is contemplating the preparation and publication of such a form, possibly in cooperation with the AGC.

[4] If *only* the specimen form will be acceptable, this should be clearly stated.

in the event the owner's actual damages (normally the incremental cost of awarding the second low bidder) are less than the amount of the bond, the owner is entitled to only actual damages. Similarly, if actual damages exceed the penal amount of the bond, the defaulting bidder is responsible for the balance. On the other hand, a liquidated damages bid bond is a fixed amount that the owner is entitled to recover irrespective of the amount of actual damages.

Some owners have a policy of returning the bid security of all but the three lowest bidders shortly after bid opening. Retaining the security of the second and third lowest bidders will generally not work a hardship on them except in the unusual case of the posting of a certified check or cash deposit. When such cash equivalents are provided, the owner may wish to seek advice of counsel concerning the responsibilities associated with holding negotiable instruments. Returning bid bonds is not necessary because the bond becomes moot automatically except in the case of a low bidder who declines to execute the contract.

Addenda

An addendum is a written or graphic document sent to prospective bidders prior to bid opening. Addenda are intended to clarify, revise, add to or delete from the bidding documents. Addenda become necessary for a variety of reasons, including clarifying ambiguities, answering questions, and changing bidding documents (e.g. changing bid opening place, date, and time, or modifying technical requirements of the drawings and specifications). It is important that the bidding documents require bidders to acknowledge receipt of *all* addenda with the bid. The acknowledgement ensures that all bidders are bidding on the same effort.[5]

Subcontractor Listing

Some owners require and others may desire that bids include a listing of proposed or intended subcontractors and principal equipment manufacturers. Requiring listing of subcontractors prior to bid opening is not recommended. While presumably well-intentioned and primarily intended to reduce the evils associated with bid shopping (in the case of subcontractors) or swapping inferior equipment for named equipment (in the case of suppliers), experience has shown that bidders have problems in listing names on the bid form because of last minute changes. Also, the use of such listing provisions is fraught with potential for disputes, even with carefully drafted contract language. Another potential concern is the problem created in which the contractor becomes obligated to use the listed party through the last-minute efforts to submit a competitive bid when a much more suitable subcontract and/or subcontractor selection could have been negotiated by the contractor in the open market after the prime contract had been awarded.

It is difficult to show a tangible benefit to the owner or contractor by requiring a subcontractor listing with the bid.

It is usually preferable to require the low bidder to identify certain subcontractors and suppliers *after* bids are opened.

Also, to ensure that the prime contractor retains control of the work, it may be desirable to contractually limit the amount of work which may be subcontracted.

Restrictive Specifications

The specifications for materials or equipment based on or naming a specific manufacturer's product can lead to difficulty if not appropriately handled. (A description of the various types of proprietary specifications and their advantages and disadvantages is beyond the scope of this publication. One useful reference is the chapter on the subject in the *Manual of Practice* of the Construction Specifications Institute.) The bidding requirements and contract documents should be specific with respect to whether substitutes "or equals" are permissible, and if so the applicable criteria and procedures. Many specification writers believe that the specifications should permit the use of equals. The salient characteristics important to the owner and engineer for the designated product or equipment should be set forth in the specifications. In this regard the coordinated series of EJCDC documents is quite useful. *See*, e.g., ¶ 9 Guide to the Preparation of Instructions to Bidders, and ¶¶ 6.7.1-6.7.3 of the General Conditions.

Preprinted General Conditions

The General Conditions found in the contract documents should be the printed rather than the typed version. Any alterations to the General Conditions should be addressed by Supplementary Conditions or annotations on the printed version. This facilitates a contractor's assessment of rights and responsibilities, and eliminates the need to review detailed General Conditions for every bid. Burying modifications in an innocuous appearing set of documents is counter to the goal of obtaining informed bids at competitive prices.

[5]Failure to acknowledge addenda normally renders a bid nonresponsive. See p. 13 below.

CHAPTER III

Recommended Procedures—Prior to Bid Opening

Selection of Prospective Bidders

One of the rudiments of competitive bidding is a sufficient number of capable bidders who are willing to invest time and money to prepare a bid and compete for the award of the contract. Limitations on who may obtain bidding documents are generally prohibited in the public sector. It is recognized that private owners may limit bidding to those contractors that they or their engineer select. Prequalification (to be discussed below) may be permissible. However, prequalification does not guarantee that the best or most competent contractors will be selected to compete for a project and it adds a separate step to the contracting process.

Where public agencies are involved, it is important that nothing be done that would create even the appearance of eliminating a qualified bidder from the opportunity to compete.

Number of Bidders

Minimum competition obviously requires at least two bona fide bids. However, it is well-established that the greater the number of bidders, the lower the bids. Thus, efforts should be made to widely advertise the project and time the bid opening so as to maximize the opportunity for bidders to prepare their bids.

Procedures for Advertising for Bids

Many owners and engineering firms maintain lists of prospective bidders and mail bidders a copy of the invitation. The invitation is also routinely distributed to accredited plan rooms and publications that specialize in compiling and disseminating information on the planning, design and construction status of projects. Public projects in some states may also need to be advertised in a particular format, e.g. legal notice, in a specified publication, e.g. the official newspaper of a municipality and for a specific time period prior to receipt of bids. The applicable statute or regulation should be consulted to ensure that the publication is sufficient as to form, content, frequency, and lead time.

Time for Preparation of Bids

Preparation of a construction bid is an expensive and time-consuming task. Owners should allow ample time for quantity take-off, site investigation, receipt of subbids, and preparation of the estimate. After all, preparing the bid for a single project is not the only thing on a bidder's agenda at a given time. Some owners and design professionals find it helpful to informally poll some prospective bidders regarding a realistic bidding period. The time required for bid evaluation and award should also be considered.

Except in emergency circumstances, a minimum of ten days should be allowed for a small, simple project. Large complex projects should allow thirty or more days. When prior approval of products is specified, the length of time should be further increased.

Bid Opening Date, Time, and Place

Since much of the work in bringing together the final numbers takes place the day and evening before bid opening, it is to the owner's advantage to open bids in the afternoon or, as a less desirable alternative, in early evening. Bid openings on weekends, Mondays, Friday afternoons, or days immediately preceding or following a recognized holiday should be avoided. Bid closings should be between 1:30 P.M. and 4:30 P.M., local time, Tuesday through Thursday. This allows contractors to work with the schedule of multiple sub-bidders necessary to prepare an informed and competitive bid. When the bid opening is to be public, it is to the owner's advantage to open bids in a place reasonably convenient to bidders. The location should be clearly identified in the bid documents and the location thereafter should not be changed unless absolutely necessary. Since many last minute bids are received from subcontractors, it is to the owner's advantage in obtaining the lowest possible price, to provide the bidders access to telephones.

Awarding authorities should contact local plan rooms or contractor associations prior to setting a bid opening date. Better competition is obtained when several major projects are not bid the same day or close to each other.

Issuance of Bidding Documents

The advertisement or invitation for bids should indicate the arrangements for viewing and obtaining bidding documents. It is advisable to issue all contract documents from a single office so that a complete and up-to-date list of all plan holders can be maintained for timely and complete issuance of addenda. Bidding documents should be available for viewing at accessible locations.

Bidders should not be charged a non-refundable deposit for a single complete set. This will discourage bidding and thus lessen competition. Additional sets should be provided at cost. Issuing authorities may also charge a reasonable handling and mailing charge for bidding documents not picked up by bidders. In addition, if multiple contracts are to be awarded, it is advisable to furnish all bidders with at least one set of the drawings and specifications for each of the contracts. Bidders should not be required to make their own documents from reproducible originals. A reasonable deposit or deposit card may be required to ensure the return of the documents. Wherever possible the non-cash system is preferable.

There should always be adequate numbers of bidding documents to ensure reasonable availability to prospective bidders on a timely basis. When bidding time is limited, the project is complicated, or there is multiplicity of subtrades involved, it is in the owner's interest to increase the number of sets of documents available. Partial sets of bidding documents should not be issued and prospective bidders should be warned that failure to review a complete set of documents will be no excuse for noncompliance.

Each bidder should be allowed to retain these sets until the contract has been awarded, or until it is definitely out of the competition, after which the documents should be returned promptly.

If a deposit is required, the deposit should be refunded upon return in good condition of the bidding documents within a specified number of days after award of the contract. If documents are not returned in good condition, the bidder's deposit may be wholly or partially forfeited.

Accredited plan rooms should be supplied with two or more sets of the bidding documents. This will increase bidding opportunities for subcontractors and suppliers. This procedure will assist in a reduction of the number of bidding documents required.

A list of document holders should be prepared and distributed to all document holders, plan rooms, trade publications and known prospective sub-bidders at least five days prior to bid opening.

Prebid Inquiries

Inquiries and requests for clarification regarding the bidding documents can pose a dilemma for the owner or engineer. On the one hand, bidders frequently do have legitimate questions or needs for clarification. The documents should require that all inquiries be directed to one individual. If any change to the documents is deemed necessary as a result of such request, it should be made by addendum. Thus the real difficulty arises when the request for clarification is not received until a short time before the scheduled date and time for bid openings. The situation may necessitate the postponement of the bid opening date. There should be a response to all questions received.

It is imperative that any answers to questions or clarifications be provided only by addenda and that the bidding documents clearly state that oral statements are not deemed to be changes in the documents.

In such a situation, the engineer needs to judge the risks involved in not giving a clarification as compared to the costs and difficulties that would flow from extending the bid opening date. In making this decision, it should be kept in mind that if a later dispute arises a contractor will be entitled to rely on its own interpretation of any ambiguity for which clarification was sought and not received.

Issuance of Addenda

Addenda should be issued to all document holders of record in an expedited fashion. If an addendum is necessary within four working days of bid opening, a decision should be made as to whether all document holders of record should be notified of the impending addendum by telephone or if the bid opening should be postponed. While postponement of a bid opening is sometimes disfavored, the benefits to be derived by insuring that all bidders are bidding on the same effort sometimes justifies a brief postponement.

An objective standard should be applied. For example, if an error or omission or other factor could justify the issuance of an addendum twenty days before the bid opening date, that same factor discovered three days before the bid opening date should justify the postponement of the bid opening. Each bidder should be required to list all addenda it has received.

Prebid Conferences

Prebid conferences are useful to familiarize prospective bidders with the site and scope of the work, particularly on larger projects. Additionally, a prebid conference affords prospective bidders an opportunity to get answers to their questions and seek clarifications, which, if not provided, could have a significant effect on the difficulty and cost of the work. However, because many bidders feel that certain questions might reveal their bidding or construction strategies to competitors, some prebid conferences generate few questions.

At prebid conferences, questions or inquiries should be reduced to writing for clarity of understanding and to aid in preparation of addenda which provides the response.

A meeting with the owner's personnel is advisable prior to the conference to determine the procedures to be followed at the meeting.

In spite of the instruction which is almost always given advising bidders that no oral statements will be considered to be an interpretation of the contract documents, the admission of this type of evidence by tribunals is widespread. It is advisable, therefore, that no oral statements be made at the conference which cannot be appropriately reduced to written form in the subsequent addenda.

Site Investigation Requirements

Recommended procedure is to require bidders to investigate the site. However, it should be recognized that the site investigation requirements have been widely interpreted to require only a reasonable visual examination, not an exhaustive exploration.

When such requirements are imposed, it is incumbent upon the owner to make appropriate arrangements relating to security, production schedules, or transportation to make the site accessible.

Withdrawal of Bid

As a general proposition, a bid may be returned on request provided the time for bid receipt has not passed and bids have not been opened. It is recommended that the owner return such bids only upon written request of one known to be an authorized agent of the bidding firm.

A few jurisdictions do not allow the resubmission of a bid by a company which has withdrawn its bid. The better practice is to permit a bidder to modify its bid prior to bid opening if it intends to resubmit a bid.

Bid Modification

It is generally permissible to allow bidders to modify their bids prior to the bid submission and bid opening times. Bid modifications should always be in writing and should be made in the same manner as permitted for the bid submission itself.

CHAPTER IV

RECOMMENDED PROCEDURES—BID OPENING

A public bid opening is one of the fundamentals of the competitive bidding process. The time and place of bid opening are set in the invitation and instructions. Individuals involved in receipt and opening of bids should exercise care to ensure that the process described in the instructions is followed.

Bids should be submitted in sealed envelopes. In no circumstances should sealed bid envelopes be opened prior to the bid opening time.

Late Bids and Late Modifications to Bids

Late bids should be treated as though they were never received. Accordingly, they should be returned unopened. Strict and rigid enforcement of this rule maintains and enhances the integrity of the process. It also eliminates the need for the owner or engineer to become involved in deciding whether a bidder had good cause for submitting a late bid.

While many bids are hand-carried and personally deposited, some contractors may use mail or messenger services to deliver bids. To help ensure that such bids are timely received, explicit directions for the place of receipt should be included in the invitation and instructions. All bids should be date and time stamped or otherwise marked when they are rceeived.

Opening Bids

Publicly-funded work typically requires that the bids be publicly opened and read aloud. The individual opening the bids should be familiar with just what aspects of the bid need to be read aloud. This can be best determined by reviewing the form to be used for a tabulation of the bids received. (Since many of the bidders or their representatives may be present, it is a courtesy to provide them with blank copies of the bid tabulation form for purposes of their note taking during the bid opening.) On a large unit price contract with many bid items the reading of one bid could be very lengthy. In such cases it may be practical to initially read only the apparent or unofficial total in each bid and, then, on the basis of these totals, read the unit prices of the first, second and third apparent low bidders.

The bid opening official should announce that the time for receipt of bids has passed and that the bids will thereby be opened. The bids should be selected from the bid box at random, opened, checked for compliance in respect of bid security, irregularities and required enclosures and then read aloud.

The originals of the bids may be made available for inspection after the opening has concluded. However, such inspection should be conducted only in the presence of the bid opening official. On the one hand, an individual may wish to examine competing bids for their responsiveness. On the other hand, the bid and accompanying information needs to be safeguarded for later evaluation.

Other matters to be considered at bid opening:

1. Presence and amount of bid security;

2. Proper acknowledgment of receipt of all addenda; and

3. Presence of any other documentation (e.g., bidder's qualifications form) required to be submitted with a bid.

When bids are not opened publicly, an owner is well advised to make a timely and public disclosure of the apparent successful bidder. Contractors are reluctant to have their bids used to bargain down an undisclosed apparent successful bidder and may be reluctant to bid to that owner again if they feel that they have been used for free estimating services. A timely, clear decision is in the owner's best interest. In either event, any non-bond security, e.g., certified check or securities, should be safeguarded by the owner.

CHAPTER V

Recommended Procedures—After Bid Opening

Introduction

After all the bids have been opened and preliminarily tabulated, the owner and its engineer should evaluate the bids and the bidders to make certain decisions with respect to award of the contract.

Bidder Evaluation—Responsibility

No owner needs to award a contract to any bidder who is not technically and financially competent. The award of a contract to a bidder based on lowest evaluated price *alone* can be false economy if there is subsequent default, delayed performance, or other unsatisfactory performance resulting in additional costs. While it is important that a public agency procure construction at a low price, this does not require award to any contractor solely because it submits the lowest bid, and the laws of most states recognize this.

The phrase "responsible bidder" refers to something more than just financial capacity of the bidder. Such factors as competence, judgment, skill, and integrity play important parts in the overall determination. The owner should determine that the bidder:

—has adequate financial resources, or the ability to secure such resources;

—thas the necessary experience, organization and technical qualifications, and has or can acquire, the necessary equipment to perform the proposed contract;

—is able to comply with the required performance schedule or completion date, taking into account all existing commitments;

—has a satisfactory record of performance, integrity, judgment and skills.

A prospective contractor may be required to affirmatively demonstrate its responsibility, including perhaps that of its proposed subcontractors. The amount of work to be done by the contractor should be examined to verify that the contractor meets the requirements of the bidding documents with respect to the amount of work to be performed with the contractors own forces.

Any determination of non-responsibility should be made by the *owner* on the basis of documented information and with advice of counsel. Under no circumstances should the engineer become involved in evaluating financial statements.

Bid Evaluation—Responsiveness

It is essential to demand conformity with all material conditions of the invitation. Failure to do so results in decreased integrity of the process. These questions frequently pit a contractor who seeks "two bites at the apple" against the engineer and owner who seek some latitude in selecting the successful bidder.

A responsive bid is one that complies in all material respects with the terms of the invitation. Each situation naturally turns on its own facts. The basic test to distinguish between a nonresponsive bid and one containing a waivable minor informality is: Are price, quality, quantity, or time affected? Can a binding contract be formed without supplementation of the bid? A responsive bid is therefore a prerequisite to the award of a contract. A nonresponsive bid (offer), i.e., one not complying in all material respects with the invitation for bids, cannot be considered for award (acceptance) and must be rejected. This compliance relates to both the method/timeliness of bid submission and the substance of the bid. For example, an unsigned bid, or a bid without a required bid bond are clearly nonresponsive.

The primary and underlying purpose behind the insistence on responsive bids is that all bidders must stand on an equal footing so that the integrity of the competitive bidding system may be maintained. A public owner's waiver of a nonconformity in a low bid will often result in litigation.

Mistakes; Correction and Withdrawal of Bids

If, after bids are opened, the low bidder claims a serious and honest error in bid preparation, and can support such claim with evidence satisfactory to the owner and engineer, withdrawal of the bid should be permitted, subject to the requirements of applicable laws. Any bid guarantee should be returned. Action on remaining bids should proceed as though the withdrawn bid had not been received.

After bid opening, a bidder should not be permitted to alter a bid and resubmit it based on a claim of error, or otherwise. Court decisions in some states have permitted correction in certain circumstances.

Dealing With an Unusually Low Bid

If one bid seems unusually low, say more than ten to fifteen percent below the nearest competing bid, it is a good practice to ask the bidder to verify its bid. Many times the bidder will confirm that it is ready, willing and able to do the project for the bid price. However, a bidder may also sometimes find a mistake and be able to establish that it is entitled to withdrawal.

Awarding to an unusually low bidder without seeking verification is usually not the bargain it may initially appear to be. If the bidder does not have enough money in the bid to do the job properly, there may be incentive to skimp or otherwise cut corners. In some instances, the bidder may begin performance but end up defaulting. On occasion the courts have refused to enforce such contracts on the theory that the owner was taking advantage of an unconscionable or unfair situation.

Determining the Lowest Bid

The instructions should set forth the basis for evaluating the bids for award purposes. Particular emphasis needs to be given to the situation where there are many bid items in a unit price contract. This situation can be even more complicated when there are combination and alternate bids. The instructions and bid form should set forth in clear and objective fashion the methodology to be used for determining the lowest bid. It is to avoid problems in this stage that the bidding documents should state the rules for dealing with obvious computational errors on the face of the bid form, as well as dealing with the ambiguity created when the arabic figures disagree with the sums written in words.

Although it is best practice to award the contract to the lowest responsive, responsible bidder on the basis of the base bid, in some instances, alternative prices are considered in determining the low bid.

An excessive number of alternates greatly increases the problems of bidders in preparing bids and can contribute to increased possibility of error. More informed and error-free bids will be obtained with a minimum of alternatives. When used, alternatives should be clear and concise.

Selection of alternatives should be made by the owner acting upon the engineer's recommendations for the best interest of the project in line with available funds. The selection should not be manipulated to favor any one bidder. Advance establishment of sequence of consideration is preferable and may be mandatory on public work.

The bidding documents should contain clear and unmistakable language describing how alternatives will affect the contract award. Volunteered or optional alternatives should not be a factor in determining the low bidder. Whenever possible, the base bid should include the entire project as the owner wishes it constructed.

Unbalanced Bids

An "unbalanced bid" is one involving multiple bid items in which the prices bid by a bidder for certain bid items are abnormally high or low. The typical unbalanced bid is the situation where a bidder believes the estimated quantity of certain bid items to be too high and the estimated quantity of other bid items to be too low. In the latter situation, the bidder may bid a high price that is expected to be applied to the anticipated overrun. To compensate for this potentially profitable move, the bidder bids a less-than-cost price on those bid items where there is an anticipated underrun. Thus, if the bidder is correct, even though it will lose money on the latter category of items, it will be more than made up on the overruns.

A variation on unbalanced bids is the technique of "front end loading", in which bid prices applied to work to be done early in the project are inflated to maximize cash flow. Again, so that the total bid is competitive, prices for work on the latter part of a project are deflated and may be less than the cost of performance.

From a contractual and bidding procedures perspective, it is difficult to develop criteria for rejecting an obviously unbalanced bid as being nonresponsive. However, the situation can be controlled to some extent on unit price contracts by including provisions in the contract documents concerning variations in estimated quantities. Unless a bid is so materially unbalanced as to prejudice the owner, it should not be rejected.

Rejection of All Bids—When Permitted

The owner normally reserves the right to reject any or all bids, typically for good and sufficient cause. The exercise of this right should be restricted to unusual circumstances. For example, all bids might be rejected if it turns out that the owner is materially underfinanced or if it is discovered that there are serious flaws in the work requirements set forth in the plans and specifications. Bids should never be rejected as subterfuge to accept a bidder who did not submit a proposal until after bids of others are submitted and made known to it; or to obtain an estimate of costs of the project so as to award it in separate contracts; or to award to a bidder selected in advance irrespective of the bidding process.

When, for good cause, bids received in competition are not acceptable and must be rejected, extreme care should be taken to avoid any requirement in rebidding the project which will compromise the principles of competitive bidding.

Reasons for rebidding a project will differ in each instance. It is impossible to lay down a guide with specific application to every circumstance. Nevertheless, it should be understood that if competitive bidding is to produce lowest possible prices for owners consistent with design requirements, bidders must have assurance that bids will not subsequently be subjected to rebidding.

If all original bids must be rejected for reasons of cost, new bids should not be solicited until the owner has sufficient time to consider the probable effects of (1) different cost factors in labor, material or equipment, or (2) design changes to bring project cost within funds available. Where redesign takes place, the changes in project requirements should be extensive enough to affect a substantial change, at least ten percent or more, in the bids.

If only one bid is received, it is the owner's obligation (1) to return the unopened bid to the bidder or (2) to be prepared to make an award on the basis of the single bid received, if the bid is not over a previously stated estimate or available funds.

Limitations on Negotiating With the Low Bidder

Sometimes following a public opening of sealed bids, it is revealed that the lowest bid exceeds either the engineer's estimate, or the funds the owner has available for the project. Assuming that the owner needs to proceed, several options are apparent. The owner may secure additional funding, or reject all bids received and readvertise (presumably with a reduced scope), or, negotiate with the low bidder to delete certain work from the contract.

Because the latter option often appears to be the most expedient, many owners and engineers will seek to use it. However, the use of such a procedure should be restricted to unusual circumstances.

One important question at this point is whether such a procedure is legal for public work. The general rule is "no", but as with most rules, this one, too, has its exceptions.

The courts in this country have quite consistently required strict compliance with competitive bidding laws. Some have stated it in terms of the principle that the contract awarded must be the same contract that was advertised. Others have said that each competing bidder must have an opportunity to submit its best price for the contract and that such bidding must be on totally equal terms. Negotiating work out of the job with the low bidder violates this principle, as the other bidders don't get the opportunity to give their price for the reduced scope.

Sanctions for violating this principle can be severe. For example, courts have held that when the contract awarded does not conform exactly to the contract advertised, such contract was formed in violation of competitive bidding laws and the contract is therefore illegal and void. Such a finding has resulted in contractors being required to repay all funds received from the owner. Likewise, in some instances, competitive bidders may have a claim for recovery of lost profits or bid preparation expenses as a result of the owners breach of the obligation to consider all bids in a fair and impartial manner.

Private owners, of course, are not so constrained.

CHAPTER VI

Recommended Procedures—Contract Award

The Firm Bid Rule

As a general rule, a contractor's bid is an offer to enter into the specified contract for the bid price. However, this offer is usually open for only a fixed time. Typically, the instructions to bidders stipulate the length of time the bid is considered firm. It should be kept in mind that it is important that the award be made before the bid expires or the contractor cannot be bound to perform. When it is not possible to make the award within the stipulated time it is necessary to obtain a written extension from the low bidder (or lowest two or three). This frequently is necessary where another organization (e.g., EPA or state) must review and approve an award, and where regulations set a relatively short period for bids to remain open.

Engineer's Recommendation

Often the owner relies on the engineer to make a recommendation concerning contract award. Engineer's evaluation and recommendation should be consistent with the applicable legal rules and the principles discussed in Chapter V above. A written narrative setting forth the basis for the recommendation should be prepared. This should address those areas where the engineer is qualified by experience and training to render judgments, leaving to other consultants and professionals the evaluation of items such as financial statements, compliance with legal requirements and adequacy of proposed insurance programs. The engineer should ensure that all information—positive or negative is true and accurate.

Legal Review

Since a determination of a given bidder as "lowest", "responsive" and "responsible" involves legal conclusions, the owner's legal counsel should review the proposed award recommendations prior to acceptance of the recommendations by the owner.

Dealing With Protests and Lawsuits

Because of the rigidity and number of applicable rules, as well as the highly competitive nature of the construction business, disappointed bidders may occasionally seek to have a given award determination changed. Depending on the jurisdiction, they may go immediately into court or they may utilize some sort of administrative protest procedure. While generalization is not always applicable, it is usually fair to say that a well-documented award determination with a rational basis and one that has had the benefit of prior legal review, is in most instances going to be sustained. The courts are reluctant to interfere with discretionary decisions of public officials.

Notice of Award

In some states, the contract will be deemed legally enforceable by both parties at the time the notice of award is dispatched to the contractor. There have even been situations where oral notice of the award has been deemed sufficient.

EJCDC Document 1910-22 is intended to serve as a guide for preparing a formal *Notice of Award.*

Contractor Response to Notice of Award

The Notice of Award will normally transmit multiple executed copies of the contract documents. The Notice of Award also normally gives the contractor a certain period of time within which to execute the documents and return them along with executed bonds and certificate of insurance (or copies of policies) and other documentation that is necessary prior to actual commencement of the work.

Executed Counterparts

Upon return of contracts signed by the contractor, they should be signed by the owner, and a fully signed copy should be returned to the contractor.

Federal and State Requirements

Award of contracts for projects funded by federal or state grant or loan funds often may require approval or endorsement of the funding agency prior to award.

Notice to Proceed

Under the EJCDC series of documents, the contract time (which is specified in the agreement) begins to run 30 days after the effective date of the agreement or on the day indicated in the Notice to Proceed but in no case later than 75 days after bid opening. EJCDC Document 1910-23 was developed as a guide for preparing a *Notice to Proceed.*

Appendix

Owners/Engineers Checklist of Bidding Procedures

	Responsible Party	Date of Completion
Prepare Legal Notice, Invitation or Advertisement		
Project identification		
Description of work		
Time of completion		
Date, time, place of bid receipt/opening		
Document availability bid security requirements		
Other statements required by law		
Prequalification requirement		
Prepare Instructions to Bidders [Ref. EJCDC Doc. 1910-12]		
1. Defined Terms		
2. Copies of Bidding Documents		
3. Qualifications of Bidders		
4. Examination of Contract Documents and Site		
5. Interpretations and Addenda		
6. Bid Security		
7. Contract Time		
8. Liquidated Damages		
9. Substitute or "Or-Equal" Items		
10. Subcontractors, Suppliers and Others		
11. Bid Form		
12. Submission of Bids		
13. Modification and Withdrawal of Bids		
14. Opening of Bids		
15. Bids to Remain Subject to Acceptance		
16. Award of Contract		
17. Contract Security		
18. Signing of Agreement		
Prepare Bid Form [Ref. EJCDC Doc. 1910-18]		
Distribute Advertisement/Invitation		
Compile bidders list		
Plan rooms		
Newspapers		
Trade publications		
Trade services		
Trade associations		
Legal notice requirements		

Arrange for Distribution of Documents

Plan rooms designated	__________	__________
Deposit determined	__________	__________
Document holders list prepared and maintained	__________	__________

Prebid Contact with Bidders

Prebid conference	__________	__________
Site available for visit/investigation	__________	__________
Arrange for secure holding of bids	__________	__________

Bid Opening

Prepare bid tabulation form/checklist	__________	__________
Verify exact time	__________	__________

Bid/Bidder Evaluation

Determine lowest bid	__________	__________
Evaluate alternates	__________	__________
Legal review	__________	__________
Determine responsiveness	__________	__________
Determine responsibility	__________	__________
Prepare award recommendation	__________	__________

Award/Execution

Formal Action to Award	__________	__________
Prepare/transmit Notice of Award (include copies of Contract Documents)	__________	__________
Review Bonds and executed copies (Legal Review)	__________	__________
Distribute executed copies of Agreement	__________	__________
Prepare/Issue Notice to Proceed	__________	__________
Receive and Review Insurance Certificates	__________	__________

INDEX

Accounting, 93
Accounts payable, 93
Accounts receivable, 93
ACEC, 20, 100, 102
ACI, 72
Action lists, 93
Advisors, 53–54
AGC, 25, 56, 57, 59, 61, 102
Agreements. *See* Contracts. *See* Short-form agreements. *See* Standard-form agreements
AIA, 20, 25, 57, 60–61, 100
American Association of Cost Engineers, 44
American Concrete Institute. *See* ACI
American Consulting Engineers Council. *See* ACEC
American Institute of Architects. *See* AIA
American National Standards Institute. *See* ANSI
American Society for Testing and Materials. *See* ASTM
American Society of Civil Engineers. *See* ASCE
ANSI, 71
APWA, 61
Architectural projects, 36
ASCE, 20, 100
ASCE Manual No. 45, 22, 23, 24
ASFE, 100
Associated General Contractors of America. *See* AGC
Association of Engineering Firms in the Geosciences. *See* ASFE
ASTM, 71

Bid evaluation, 86–87, 93
Bid opening, 57–58
Bid solicitation, 57, 93
Bidding, 21–22, 55, 56–58, 85, 86–87; role of design professional, 56; procedures for public work, 57–58
Bonds, 104
Bonus clauses, 75
Budget planning, 80

Cash-flow requirements, 73, 76, 93
Certificates of completion, 77
Change orders, 73, 75, 93
CMAA, 61
Codes and standards, 44–45, 53
Communication, 6–7, 10, 16–18, 64–65, 77; critical points in, 17–18; during design, 32; forms of, 16; importance of, 14; timing, 18
Competitive bidding. *See* Bidding
Computer applications, 63, 89, 92–94
Computer hardware, 90
Computer software, 63, 90
Computer systems, 90
Computer-aided design (CAD), 42, 94
Computerization, 89–91
Conceptual design, 91
Conflict avoidance, 105
Conflict resolution, 18, 105–106
Conflicts, 6
Connections, 68
Constructability reviews, 44
Construction contract documents, 60
Construction contracts, content of, 60; form of, 60–61; international, 61; standardization of, 61. *See also* Contracts
Construction costs, 43–44
Construction documents, 85
Construction facilities, 64
Construction Management Association of America. *See* CMAA
Construction phase, 80–81, 85–86
Construction supervisor, 52–53
Construction support services, 53
Construction team, 11–12, 51; assembling, 54
Constructor, 6, 11–12; insurance needs of, 105; qualifications of, 55–56; quality assurance/quality control program, 87–88; requirements of, 2; responsibilities, 12, 62–63, 65, 70, 81; selecting, 55, 56–59
Consulting Engineering: A Guide for the Engagement of Engineering Services. See ASCE Manual No. 45
Contract award, 58
Contract documentation, 67–68
Contracts, 23–25, 38, 49–50, 51, 87, 103–104; negotiated, 58–59. *See also* Construction contracts. *See also* Short-form agreements. *See also* Standard-form agreements
Control strategies, 80
Coordination, 64–65; during design, 32; team members' roles, 14–15
Coordination process, development of, 15–16
Cost, 28; control, 64; estimates, 64, 73, 93. *See also* Construction costs. *See also* Design costs
Cost-plus contracts, 74

Data retention and retrieval, 90–91, 94
Design, computer applications, 91–92
Design activities and requirements, 42–44
Design considerations, 43
Design costs, 11, 28, 32–33
Design guidelines, 32
Design phase, 79–80, 85
Design plan, 31
Design professional, 6–7, 11, 39–40, 53, 56, 84–87; compensation for services, 21, 24; during construction, 39; insurance needs of 104–105; qualifications of, 19; quality assurance/quality

control program, 84–87; relationships with owner and constructor, 42; requirements of, 2; responsibilities, 11, 45–46, 62, 65, 70, 86–87; selecting, 19–22; services, 23–24
Design quality, 38
Design responsibility, 45–46
Design reviews, 43, 86
Design team, 11, 32, 35–39
Design team leader, 31–32
Design-construct project, 36
Documents, filing and storing, 41–42, 99. *See also* Records
Drafting practices, 42. *See also* Computer-aided drafting (CAD)

EJCDC, 25, 57, 59, 60–61
EJCDC documents, 23
Engineering project, 35–36
Engineers Joint Contract Documents Committee. *See* EJCDC
Environmental impacts, 28–29, 80
Equipment, 69

Fast-track construction, 33
FIDIC, 61
Field organization for construction, 51–54
Final design, 92
Financial resources, 47
First aid, 64
Forms and agreements, 94

General ledger, 93
Government grants, 45
Governmental agencies, 25. *See also* Regulatory agencies

Human resources, 48

Incentive clauses. *See* Bonus clauses
Income property analysis, 94
Insurance, 104–105
International federation of Consulting Engineers. *See* FIDIC
Interviews, 20–21, 99
Invitation for proposal, 20

Job-cost reporting, 93

Library, reference, 42
Life-cycle cost, 11, 43, 80
Liquidated damages, 75
Litigation, 106
Lump-sum contracts, 64, 74

Manufacturing capability, 48
Materials, 47–48, 71–72; management, 64, 94; substitution of, 72–73
Mechanical and electrical components, 69
Meetings, 17, 32, 63
Minimum acceptable standards, 72
Monitoring, 32
Multidiscipline projects, 35

National Society of Professional Engineers. *See* NCEC
Negotiating phase, 85
Negotiating process, 86–87
Network scheduling logic, 93
NSPE, 20

Office operation, 41–42
Operating phase, 82–83
Operation and maintenance, planning, 79–81
Organizational levels, 35
Organizational peer review, 97, 98, 99
Owner, 10–11; expectations, 5, 6; insurance needs of, 104; requirements of, 2, 5–6; responsibilities, 5, 10, 24, 51, 54, 65, 70, 81–82; role, 5
Owner's selection committee, 19
Owner's team, 10–11

Payments, 73–74
Payroll, 93
Peer reviews, 44, 95–97; benefits of, 97; elements of, 98–100; reports, 99; request for 98; scope of, 98–99; types of, 97–98
Peer review programs, 100
Permits, 45
Personnel, 80
Physical plant, 80
Placing drawings, 69
Pre-bid conference, 57, 86
Pre-project phase, 85
Preconstruction meetings, 63
Preliminary design, 79, 91–92
Private work, competitive bidding for, 58
Professional disciplines, 35, 39–40
Progress reports, 32, 76–77
Project construction, organization for, 62–63
Project impacts, 28–29
Project management, 93–94; peer reviews, 97, 100
Project manager, 8, 10
Project peer reviews, 44, 97, 98, 100
Project performance peer review, 98, 99
Project phases, 85–86
Project planning, 8, 10
Project programming, 91
Project requirements, 35–39; alternatives, 27–28, 29–30; refining, 27
Project team, composition of, 2; organization of, 8; performance of, 102–104. *See also* Team members
Project-specific requirements, 87–88
Public influence, 29
Public owners, 6
Public work, 73; competitive bidding for, 56–58

Quality, definition, 1–2; requirements for, 2–3; threats to, avoiding, 33

Quality assurance, definition, 84
Quality control, definition, 84; design-related, 44
Quality takeoff, 93

Records, 77. *See also* Documents, filing and storing
Regulatory agencies, 2, 45, 48–49, 53, 68
Reinforcing steel components, 69
Reports, peer review, 99–100
Requisitioning, 93
Resident project representative (RPR), 53, 62, 72, 73, 74, 75, 76; qualifications, 54; responsibilities, 51–52, 71, 77–78
Retainage, 74–75
Reviewers, selection of, 99
Risks, 102–104

Safety, 64, 80
Schedules, 8, 10, 28, 32–33, 43, 63–64, 67, 76
Schematic design phase, 85
Serviceability, 43
Shop drawings, 67, 68–70, 94
Short form agreements, 25
Site development, 49
Standard-form agreements, 25
Start-up phase, 81–82
Statement of qualifications, 20
Statistical analysis, 72
Structural components, 67, 68–69
Submittals, 66–67
Suppliers, 48

Team members, 12–13, 15–16, 103; key contacts, 14; obligations of, 3; roles in coordination, 14–15. *See also* Project team
Technical documentation, 68
Telecommunications, 94
Temporary construction, 66, 70
Testing, 70
Two-envelope system, 22

Unit costs, 44
Unit-price contracts, 73–74

Value engineering, 96

Warranties, 104
Word-processing, 94
Work force management, 64
Workmanship, 72
Written communication, 16, 77